Shale Gas and Fracking
The science behind
the controversy

Shale Gas and Fracking
The science behind the controversy

MICHAEL STEPHENSON

ELSEVIER

Amsterdam • Boston • Heidelberg • London • New York • Oxford
Paris • San Diego • San Francisco • Singapore • Sydney • Tokyo

Elsevier
Radarweg 29, PO Box 211, 1000 AE Amsterdam, Netherlands
The Boulevard, Langford Lane, Kidlington, Oxford OX5 1GB, UK
225 Wyman Street, Waltham, MA 02451, USA

Notices
Knowledge and best practice in this field are constantly changing. As new research and experience broaden our understanding, changes in research methods, professional practices, or medical treatment may become necessary.

Practitioners and researchers must always rely on their own experience and knowledge in evaluating and using any information, methods, compounds, or experiments described herein. In using such information or methods they should be mindful of their own safety and the safety of others, including parties for whom they have a professional responsibility.

To the fullest extent of the law, neither the Publisher nor the authors, contributors, or editors, assume any liability for any injury and/or damage to persons or property as a matter of products liability, negligence or otherwise, or from any use or operation of any methods, products, instructions, or ideas contained in the material herein.

ISBN: 978-0-12-801606-0

British Library Cataloguing in Publication Data
A catalogue record for this book is available from the British Library

Library of Congress Cataloging-in-Publication Data
A catalog record for this book is available from the Library of Congress

For information on all Elsevier visit our
website at http://store.elsevier.com/

Printed and bound in the USA

Working together
to grow libraries in
developing countries

www.elsevier.com • www.bookaid.org

DEDICATION

To Jack and Fred

CONTENTS

PREFACE

In 1821, William Hart dug – with pick and shovel—a 27-foot deep gas well in the village of Fredonia, Chautauqua County, New York, within a few miles of the shore of Lake Erie. The well provided the light of "two good candles", but by 1825 it supplied enough natural gas for lights in two stores, two shops, and a grist mill. The pipeline to transport the gas was made from hollowed-out logs connected together with tar and rags. In 1850, the well was deepened to 50 feet and produced enough gas to light 200 burners. So the Fredonia Gas Light Company, North America's first gas company, was born and so was shale gas—though no one would have guessed the transformation that shale gas would bring almost 175 years later.

Shale gas has changed America. Within 20 years it could be providing half of the United States' domestic gas and may be being exported in a liquefied form. Shale will turn the US from energy importer to exporter, and it's likely to revitalise industry, bring hundreds or thousands of jobs, and attract back companies that long ago left America in search of cheap manufacturing costs. The questions on investors' and technologists' minds are: Will shale gas take hold outside the United States? Will technology developed in the gas fields of Texas and Pennsylvania work elsewhere – under the farmland of Europe and the steppes of Asia?

But shale gas has a darker underside. Many people believe it could be damaging the environment and diverting attention away from the need to reduce carbon emissions and ditch fossil fuels. Shale gas and hydraulic fracturing produce opinions that are clearly very polarised and politicised. So people are looking for unbiased information: investors—because they want to know if shale gas might be a safe bet; policy makers and regulators—because they want to know how to manage and control shale gas. The public is interested too, particularly those that live in prospective areas for shale gas.

But information is often thinly disguised propaganda for one or an other side of the debate. The companies that perfected the technique of hydraulic fracturing are often seen as less-than-transparent about how they make it work and their assurances that it's safe seem hollow to people who live close to fracking rigs. The environmental pressure groups serve up nightmare scenarios of earthquakes, volcanoes, and subsiding land—and creeping disease in animals and humans.

One source of information that does matter is published science, which through the process of peer review, can be said to be independent. In this book I will review and explain some of the key studies that form the background to the debate and explain the wider science and geological processes that are involved. I will look at how the amount of gas in shale is assessed, the basics of its formation, and the reasons why it's concentrated in certain places.

But shale gas is part of a wider problem to do with our use of the "underground." People worry about extraction of coal bed methane, the underground storage of gas, and the disposal of nuclear waste or CO_2 in carbon capture and storage in the same way as they do about shale. We need to see the underground as part of the environment, just as we do the land, the sea, and the atmosphere. We also need to monitor and manage the underground as well as we do the surface. Part of this book will discuss how we might do this.

ACKNOWLEDGEMENTS

Thanks to Ruth O'Dell for reading the manuscript and pointing out geologists' assumptions about the non-geological world. Thanks also to Mike Ellis, Andrew Bloodworth, David Kerridge, Rob Ward and Alex Stewart for reading parts of the manuscript. Ed Hough, John Ludden, Bob Gatliff, Nick Riley and members of the BGS shale gas team supplied lots of advice and information in the last two years, as did Toni Harvey. Even with all this expert help, however, any errors are entirely my own.

CHAPTER 1

The Fuss about Shale

Contents

When I was a student of geology in the 1980s, shale was just the useless stuff that you drilled through to get to the good stuff – rocks like limestone and sandstone that might contain oil and gas. But there's been a huge change in the world of oil and gas: the rise of unconventional hydrocarbons. I remember looking up a definition of the phrase 'unconventional hydrocarbons' a few years ago and getting the rather unhelpful '…hydrocarbons produced in a way that is not conventional…' but in fact this isn't so far from the truth. 'Unconventionals' as they are called in the industry are rather new and they involve getting oil and gas from rock layers where we couldn't get them before – or from rock layers which we never thought would yield hydrocarbons. Shale gas is the best example and it has transformed the energy landscape in the United States. Shale gas will provide half of US domestic gas production before long. US refineries are configuring themselves for gas and oil from shale. Energy is so cheap that manufacturing costs in the US are as low as some parts of China. But there's a huge fuss surrounding shale – rarely has a technique in the oil and gas industry attracted so much attention. The discussions are remarkably polarised with some people passionately supporting shale gas because of its ability to generate wealth, and others doubting that its extraction can be done safely and consistently with our wishes to reduce climate-changing carbon dioxide. In this chapter I'll examine the reasons for the controversy and set the scene for the rest of this book, looking at the science behind the issues.

Keywords: Earthquakes; Energy; Fracking; Hydraulic fracturing; Pollution; Shale gas; Unconventional hydrocarbons.

The reason why there's such a fuss about shale gas is simple. First it might provide cheap energy and change the geopolitics of energy forever. But the flip side is the fear that 'hydraulic fracturing' (nicknamed 'fracking') – which is the method of getting most of the gas out – might be damaging the environment. Let's examine the fuss on both sides. What's being said about shale gas as a 'game changer' and the environmental effects of its exploitation?

Shale gas and fracking
http://dx.doi.org/10.1016/B978-0-12-801606-0.00001-7

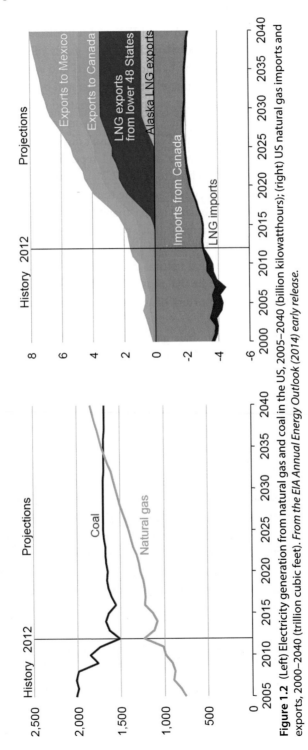

Figure 1.2 (Left) Electricity generation from natural gas and coal in the US, 2005–2040 (billion kilowatthours); (right) US natural gas imports and exports, 2000–2040 (trillion cubic feet). *From the EIA Annual Energy Outlook (2014) early release.*

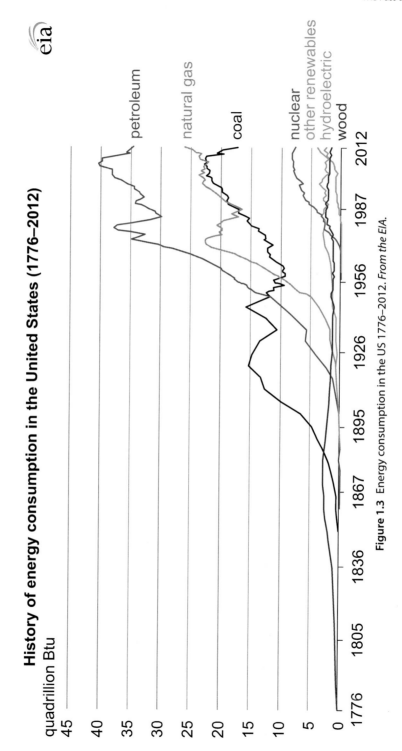

Figure 1.3 Energy consumption in the US 1776–2012. *From the EIA.*

was the US Department of Energy's (DOE) Eastern Gas Shale Project which looked at the possibilities of shale gas in Illinois and Michigan. The Gas Research Institute (GRI) was established in 1977 and in the 1980s and early 1990s it started research programs on transmission, distribution and markets of gas.

It sometimes sounds unreasonable to connect big advances to single people, but in the case of George Mitchell, this might be justified. In the 1980s and 1990s, Mitchell's company drilled 10,000 wells around Fort Worth in north Texas and spent $6 million getting the Barnett shale to release its gas. Essentially the big jump was to use practical engineering and trial and error alongside the science and technology developed by DOE and GRI programmes. Mitchell used horizontal drilling which followed the rock layers and low-cost hydraulic fracturing to make the extraction of shale gas economic. The process wasn't easy. Tests began in 1981 and production from these early wells often didn't cover drilling and operational costs.

Since 2000 drilling techniques have continued to develop and increases in oil and gas prices since 2003 have made shale gas ever more economically attractive. From the mid-1980s to 2003, the price of oil was generally under $25 a barrel, but peaked at over $140 in 2008, all increasing the ability of companies to invest in shale but also to encourage a drive for greater US oil and gas security.

Much of the early progress in shale gas was made by small companies like Mitchell Energy but in the early 2000s larger oil and gas companies entered the business, mainly by buying up the small companies. Devon Energy bought up Mitchell Energy and Exxon Mobil bought XTO. These new larger companies were responsible for much of the development of Texas shale. Between 1997 and 2009, more than 13,500 gas wells were drilled – mostly of the horizontal variety – in the Barnett shale (Fig. 1.4). In 2004, deep gas production from the Barnett shale overtook the level of shallow shale gas production from other historic shale areas such as Ohio and Michigan.

In turn the success of drilling in the Barnett shale attracted companies to other deep US shales including the Woodford (Oklahoma), Fayetteville (Arkansas), Haynesville (Louisiana), Marcellus (Pennsylvania) and Eagle Ford (Texas). Production of gas (and shale oil) from these layers is responsible for the increase shown in Fig. 1.1. The relative importance of the different shale layers is shown in Fig. 1.5.

How has increasing shale gas production affected US energy security? Policy makers studying reports in the 1980s and 1990s would have been depressed by the 'peak oil theory'. This idea, which came from academics and specialists in the oil industry, held that production of oil and gas followed a bell-shaped curve through time and that the US was well past its peak. So more energy dependence for the US was envisaged. In fact American conventional natural gas production peaked in the early 2000s in the face of steady increase in demand, and the widely discussed solution to the shortfall was to import LNG from other still-producing conventional natural gas fields in the Middle

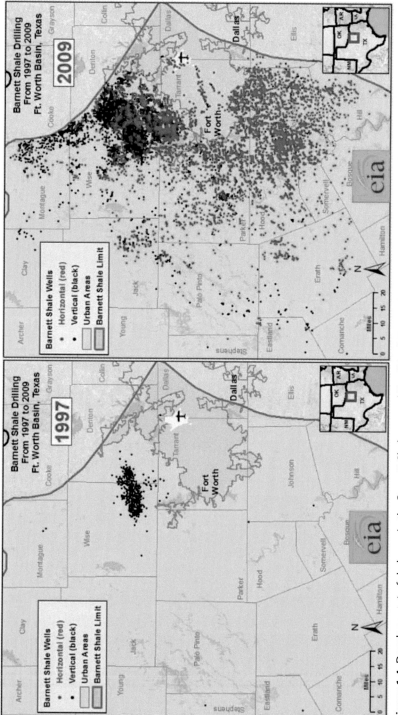

Figure 1.4 Development of shale gas in the Barnett Shale around Fort Worth Texas. The red and black dots are shale wells. *From Newell (2011).*

annual shale gas production
trillion cubic feet

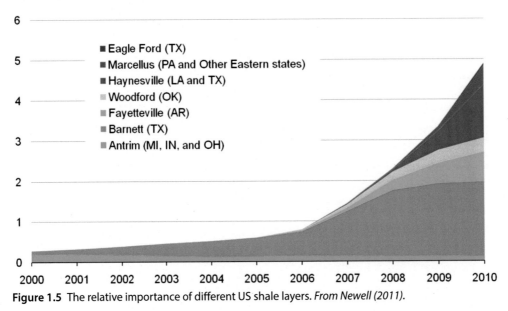

Figure 1.5 The relative importance of different US shale layers. *From Newell (2011).*

East, Australia and Russia. In 2006 there were four operating onshore LNG import ter-
minals in the US and more were expected to be built.

But shale gas (and its close relative shale oil) has changed all this. Shale gas production
has reduced US natural gas imports to a level not seen since 1994. In fact the EIA Annual
Energy Outlook 2012 suggests that the US will become a net exporter of gas in the
form of LNG in 2016. More than 2 billion cubic feet may be being exported by 2019.

What has been the effect of shale gas on US CO_2 emissions? The Kyoto protocol of
2005 was established to stabilise greenhouse gas concentrations. Amongst the industrialised countries to sign the protocol only the US did not ratify it. The irony is that
despite its reluctance to ratify – through the production of shale gas – the US appears to
be the only major industrialised country to have reduced its CO_2 emissions, mainly
through burning gas in power stations rather than coal. According to the International
Energy Agency (IEA), fossil fuel CO_2 emissions in the US fell by 430 million tonnes
between 2006 and 2011. Meanwhile *global* fossil fuel CO_2 emissions increased to a
record high of 31.6 gigatonnes (31.6 thousand million tonnes) by 2011.

US industry is being boosted by shale gas. The EIA Annual Energy Outlook 2014
predicts a jump in manufacturing powered by cheap fuel. The main industries affected will
be energy-intensive bulk chemicals and primary metals, both of which provide products
used by the mining and other downstream industries such as fabricated metals and machinery. The bulk chemicals industry is also a major user of natural gas and, increasingly, other

hydrocarbon gases which are often produced with methane from shale. One of these is ethane (a slightly larger version of the methane molecule with two carbon atoms not one) which is used to make ethylene which in turn has many industrial products including PVC, polystyrene, latex, detergent and vinyl. The Dow Chemical company announced in 2012 that it would build a $1.7 billion ethylene and propylene production facility in Texas no doubt in the belief that ethane from shale gas will be available for years to come.

The low price of US fuel now means that US manufacturing costs are lower than in coastal Chinese cities. Companies that once left the US to find cheaper production abroad are 'reshoring' or returning production to the US. These lower American energy prices could result in 1 million more manufacturing jobs as companies build new factories. Shale gas created 600,000 jobs in the US as a whole up to 2010 and is predicted to provide 1.6 million by 2035. A key reason for the shale gas industry's economic impact is the indirect jobs created to support the industry. For every direct job created in the shale gas sector, more than three indirect jobs are created, a rate higher than the financial and construction industries. Shale gas is predicted to contribute $231 billion to US GDP by 2035. In Pennsylvania alone, the home of the huge Marcellus shale, the shale gas business is predicted to bring $18 billion extra business, nearly $2 billion local tax and 200,000 jobs by 2020.

Outside the United States

All this activity has made other countries and governments interested in the possibilities of shale gas. Many of these have similar problems of security of supply, dependence and high energy prices. But where is the shale potential outside the US?

In a huge and ambitious report in 2013 that attempted to assess the size of the shale gas resource in many countries across the world, the EIA came up with some figures that showed that shale gas was not in any way confined to the US. In fact it showed that much of the world's shale gas is probably outside of the US, in Canada, China, Argentina and Mexico (Fig. 1.6).

The map shows some of these figures, contrasting estimated shale gas with already proven natural gas reserves. In several countries, the shale gas potential seems much better than conventional gas, for example in Mexico and Argentina. The interesting conclusion though is that much of this gas worldwide is in places that we don't associate with big gas resources or 'big energy'. The EIA report didn't look at Middle Eastern shale gas resources, but even if these are large (they probably are), the distribution of shale gas does rather alter the energy map of the world away from concentrations in the Middle East and Asia, to China and the Americas. These figures are enough to make the people that think about energy security look twice.

Canada is the next most developed shale gas area after the US with about 3 billion cubic feet of gas being produced per day in 2013, mainly from the Muskwa shale in the Horn River area in northeast British Columbia, the Montney shale in British Columbia

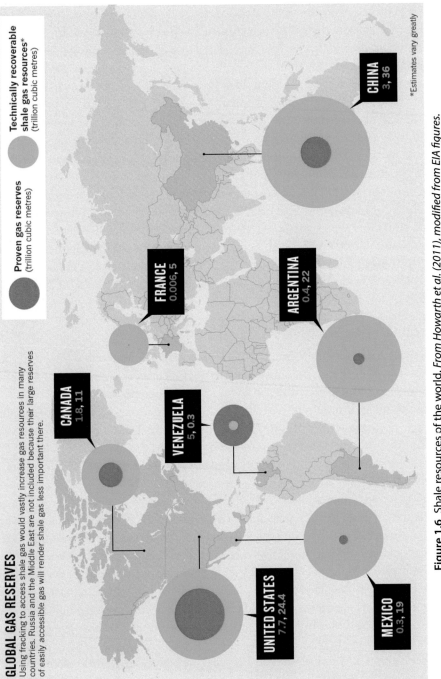

Figure 1.6 Shale resources of the world. *From Howarth et al. (2011), modified from EIA figures.*

and Alberta and the Duvernay shale in Alberta. Some of the most productive of the American shale layers also extend under the international border into Canada including the Antrim, Utica and Marcellus shales.

Of the European countries that seem to have a substantial shale resource, Poland is perhaps closest to realising a shale gas industry – and it has good reason. Poland consumes about 14.5 billion cubic metres of natural gas annually. Almost 70% of natural gas is imported; the rest is produced domestically. But Poland's energy is dominated by coal – almost 90% of its electricity is generated using local Polish Silesian coal. There are several tensions that the country has to deal with. One is that being an EU member state it is required to reduce its CO_2 emissions through the European Union Emissions Trading Scheme (ETS) which is based on 'cap and trade'. The 'cap', set by a government, is a limit to the amount of CO_2 a factory or a power station can emit. Presently the ETS enforces emissions reductions in 11,000 factories and power stations in 31 countries. Though the cost of emission is small at the moment, the 'price for carbon' is likely to rise as Europe comes out of recession and begins to generate more electricity. Poland will be hit disproportionately because of its many coal power stations. So Poland would like to switch away from coal. Not only because of the reduced CO_2 emissions but also because of the winter time smog that affects Poland's southern cities because of coal combustion.

Using gas to generate electricity would be attractive. A recent report commissioned by the EU suggested that countries like Poland could reduce their CO_2 emissions by between 41% and 49% if they switched power generation from coal to gas. But Poland has very little prospect of conventional gas. It could import gas but like much of eastern Europe it's likely that the gas would come from Russia. Poland is nervous about gas supplies from this source because of the monopoly it represents and because it thinks that supplies might be used politically.

All of these factors explain why the Polish government is so much in favour of shale gas. Licenses for exploration have greatly increased so that much of the country is now targeted (Fig. 1.7).

It also goes a long way to explaining why the people of Poland are on the whole more supportive of shale gas development than in other European countries. According to a survey by the Polish Public Opinion Research Centre in September 2011, 73% of Polish citizens are in favour of shale gas exploitation.

But progress in Poland has so far been slow. By the end of 2012, companies had drilled 33 exploration wells and fracked 11 of those. The results from drilling and testing are mixed. A few companies have revealed good results but for others, like the US giant Exxon Mobil, the results have been poor enough to prompt their exit from the country. The Polish government nevertheless thinks that commercial gas production will start in 2014, in the Wejherowo area, though some experts say that 2015 or 2016 is more likely.

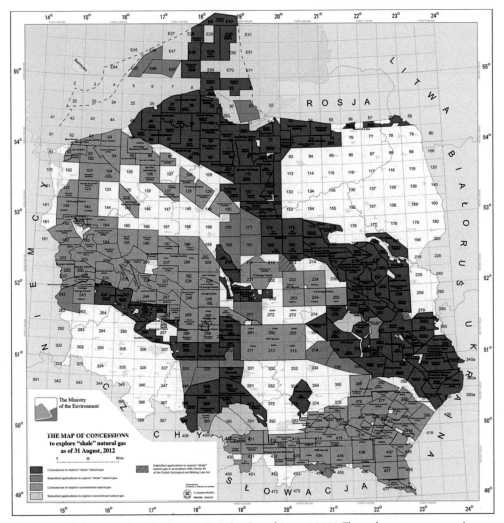

Figure 1.7 Shale gas exploration licences in Poland as of August 2012. The red areas are concessions or licence blocks where exploration for shale gas is permitted. The pink areas are where licences for shale gas exploration have been applied for. *From http://polishshalegas.pl/en/shales-in-poland/other-licensees.*

Worries about gas supply aren't just confined to Poland. Europe gets 24% of its gas from Russia, and half of that – 80 billion cubic meters (bcm) a year – passes through Ukraine. An argument between Russia and Ukraine led to the pipelines shutting down for 2 weeks in January 2009, to much consternation downstream. Events in Crimea in 2014 have increased those fears. EU countries have 36 bcm of stored gas and could store 75 bcm, but even that could run out and ultimately its top ups would mostly come from Russia. For Estonia, Latvia and Lithuania, the situation is even worse because they get all their gas from Russia (Fig. 1.8). Bulgaria gets almost all its gas from a Russian pipeline

Gas supplied by Russia, % of total 2012

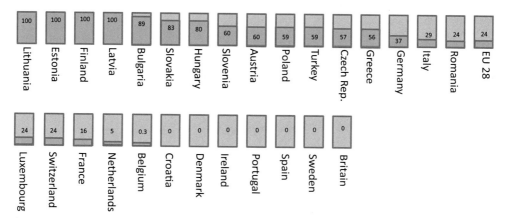

Figure 1.8 Dependence on Russian gas in Europe. *Source: Eurogas and the Economist.*

that crosses Ukraine, and it has limited storage (less than two months' consumption). So shale gas in the basins of Europe hold special fascination for policy makers at the moment and Europe appears to be well endowed with the stuff (Fig. 1.9).

In Britain, the energy security story is different. Britain has more diverse energy sources than most of Europe. If we take electricity for example, the graph (Fig. 1.10) shows that the UK gets electricity from three big sources, coal, gas and nuclear. The rise in gas looks rather like the increase in US domestic gas production as shown in Fig. 1.1, but the gas burnt in British power stations is mostly imported from Qatar (as LNG) and Norway. Like Poland, Britain is a net gas importer and 2011 was a landmark year because for the first time the country imported more methane natural gas than was pumped from Britain's offshore gas fields.

A diverse set of energy sources is a good thing, but security of supply is still a concern because as British North Sea oil and gas declines, dependency on imports will increase. Conventional oil and gas production from the North Sea passed its peak just before 2000.

Though British oil imports come mainly from Norway (Fig. 1.11), other sources are in less stable countries and there is a concern that world events like the 'Arab spring' that began in 2010, might affect Britain's ability to source its energy. Before the Arab spring, Libya supplied oil to the UK, but after the Libyan conflict imports collapsed with the result that in 2011 Libya itself was forced to import oil despite having amongst the largest oil reserves in the world.

Britain also has a very ambitious target of greenhouse gas reduction of 80% by 2050 meaning that for electricity, for example, a large amount of energy production needs more and more to be provided by low carbon sources. Nuclear and wind are perhaps the most important of these and while wind is developing well, it is expensive.

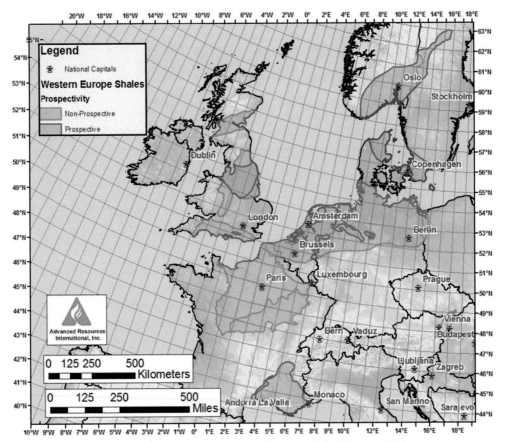

Figure 1.9 Shale gas basins in Europe. *From Energy Information Administration (EIA) 2013.*

More expensive still is nuclear. Britain has a long history of nuclear and was one of the first countries to develop commercial nuclear. However, the cost of building and decommissioning power stations – and the public dislike of them – has made their development more and more difficult. Until recently, the future for nuclear power in Britain looked bleak because many of the established old power stations are due to close (Table 1.1).

Another of the UK's plans to reduce emissions while still using fossil fuels in power stations – carbon capture and storage – has been slow to start. The technology, which has only been proven on a small scale, is expensive and difficult to commercialise without a high price for carbon emissions per tonne through the ETS.

Thus shale gas would tick many of the boxes for Britain and is therefore receiving considerable government support, including tax incentives for companies willing to drill.

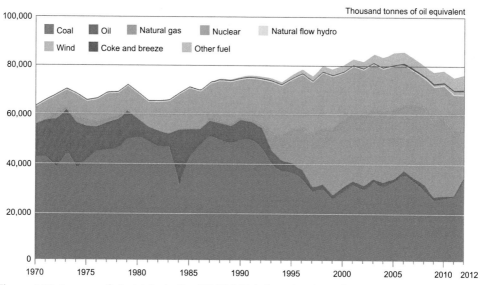

Breakdown of sources of electricity (1970–2012)

Thousand tonnes of oil equivalent

Coal Oil Natural gas Nuclear Natural flow hydro
Wind Coke and breeze Other fuel

Figure 1.10 Sources of electricity in the UK. BBC Website using DECC figures. *http://www.bbc.co.uk/news/business-24823641.*

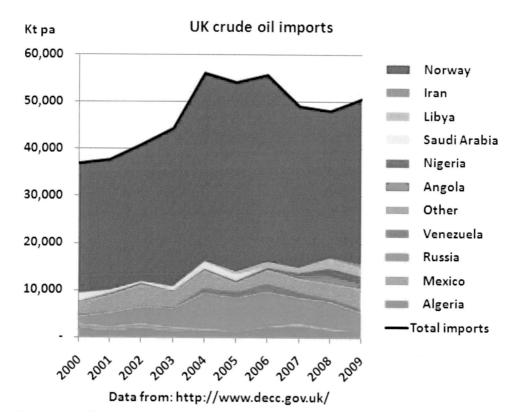

Figure 1.11 UK crude oil imports. *From http://www.crudeoilpeak.com/?page_id=2820 using DECC figures; http://www.decc.gov.uk/.*

Table 1.1 The closure of UK nuclear power stations puts pressure on the UK's ability to generate low carbon electricity

Power station	Construction started	Connected to grid	Commercial operation	Closure date
Wylfa	1963	1971	1972	2015
Dungeness B	1965	1983	1985	2018
Hinkley point B	1967	1976	1976	2023
Hunterston B	1967	1976	1976	2023
Hartlepool	1968	1983	1989	2024
Heysham 1	1970	1983	1989	2019
Heysham 2	1980	1988	1989	2023
Torness	1980	1988	1988	2023
Sizewell B	1988	1995	1995	2035

Source: DECC data.

THE FLIP SIDE

That's all good. So why do some people hate shale gas and fracking so much?

What does fracking mean to the public outside of the world of energy and policy? I've not done any systematic study but I would hazard a guess that in many people it will invoke images from the now famous *Gasland* film by Josh Fox, of householders igniting their tap water. These images have been replicated in other TV programs, for example in a British documentary in the BBC's flagship *Horizon* series that showed the geochemist Rob Jackson of Duke University igniting water from an outside tap on a farm in Pennsylvania. Jackson's team reported high levels of methane in groundwater in one of the few scientifically peer-reviewed studies of shale gas contamination in active fracking areas. They sampled wells in north-eastern Pennsylvania and found methane in 82% of them. The concentration of the gas in wells within 1 km of fracking operations was six times greater than the background level (Fig. 1.12).

Back in 1987, the US Environmental Protection Agency (EPA) reported that fracking of a well in Jackson County, West Virginia had contaminated a water well in 1984. The best documented case of water contamination is in the town of Dimock in Pennsylvania where after intensive fracking, residents complained about dirty well water that smelt of sulphur. In January 2009, a backyard water well blew up. The explosion was reported to have occurred due to methane build up, and testing of water showed high levels of methane.

People also complained of poor health. Companies were questioned about the situation and attributed it to natural methane in shallow rocks, not methane finding its way from deep below into water wells. Chemical testing of the water found chemicals associated with fracking and local people sued the operating company. Given that the company had 130 drilling violations in the area, it seems not unlikely that some contamination had occurred.

Figure 1.12 Lighting the methane in groundwater from water wells. *From © JIM LO SCALZO/epa/Corbis.*

EPA checks have since given Dimock a clean bill of health but some people worry that local landowners will sometimes underplay any damage in order not to undermine house or land value. In other parts of the state, landowners who have sued companies have settled out of court and agreed not to say any more about incidents. So it's possible that problems are underreported. Another aspect of the business that causes concerns for local residents is the so-called Halliburton loophole that excuses companies from disclosing all the materials that they use in the fracking fluids.

In February 2011, the *New York Times* reported that Pennsylvania fracking operations '…hauled wastewater to sewage plants not designed to treat it', and that it was then discharged into rivers that supply drinking water. The paper mentioned '…radioactivity at levels higher than previously known, and far higher than the level that federal regulators say is safe for these treatment plants to handle…'. Marvin Resnikoff, a consultant on radioactive waste management, lamented that rock cuttings from drilling, holding radioactive fluids, were considered solid waste and so were capable of being disposed of in county landfill sites.

It seems that even the claim that shale gas has reduced American greenhouse gas emissions can be challenged. It's all to do with methane emissions from the drilling sites. The sight of excess gas being burned in refineries and production sites isn't rare. Apart from being wasteful, this isn't a particular problem for climate change beyond the

CO_2 produced in the burning itself. The problem is the unseen methane that might escape in the early days of production from a well site when the machinery is being set up to capture gas from below. No one really knows how much methane is produced at these times. But it is well known that methane is many times more potent than CO_2 as a greenhouse gas. In a controversial study published by scientists from Cornell University, the low emissions of shale gas burned in power stations were considered to be offset by the unplanned emissions of methane (known as fugitive emissions) and the emissions associated with the energy required to get the gas out of the ground (the energy expended in fracking). They concluded that shale gas was even worse than coal for production of climate-changing greenhouse gases.

Even if we do accept that American CO_2 emissions have dropped because less coal is being burned, coal that is mined does tend to find its way to other markets. This has happened: American coal that would have been burned for power is now unwanted and therefore cheap in the US and being sold in Europe. Quite apart from the effects on the European coal industry, it shows starkly that CO_2 reductions might have been achieved in America but that the extra CO_2 has effectively been exported. It hasn't gone away.

Other commentators worry about the bad economic effects of very cheap gas. Nuclear electricity generation which has more or less zero CO_2 emissions is decreasing in the US according to the EIA Annual Energy Outlook 2014, perhaps because it's cheaper to use gas. It's probably a good thing that shale gas is displacing coal in electricity generation in the US but do we want it to displace nuclear? Shale gas is so cheap at the moment that it's hard even for the drilling companies to make money. Much of our hopes for averting climate change rest on renewable energy and many think that shale gas can form a bridge to low carbon until renewables can take the full strain of our energy-hungry society. But if gas is too cheap there'll be no incentive to make renewables work. We want gas to displace coal but we don't want it to displace renewables.

In the United States, the benefits of the 'boom and bust' hydrocarbon industry are also being questioned. Although shale gas employs people and has strong local benefits, researchers at Cornell University have pointed out that natural resource extraction industries typically play only a small role in state-wide economies and their employment impact is tiny compared to industries such as retail or health services. It's true that as drilling companies move into communities, there's more trade and more money spent in shops, bars, restaurants and businesses. New jobs are created in hotels and retail. Landowners receive royalty payments and have extra spending money in their pockets. However to understand the economic effects of shale gas drilling, it's important to understand accurately where the jobs go and how long they last. Also what are the hidden costs of shale gas drilling to the public in the form of road repairs and reductions in tourism and how will the costs and benefits be distributed? Above all how long will the boom last, and what happens when it ends?

In Europe...

For many British people fracking will almost certainly conjure a word rather than an image of burning tap water to mind. That word would be 'earthquake'. Britain has no shale gas industry to speak of, but what little it has became famous not because of flaming taps but because fracking caused a couple of very small earthquakes. (Many geologists would prefer to call them 'tremors'). On Friday, April 1, 2011, an earthquake of magnitude 2.3 occurred with an epicentre just two miles from the site of Britain's first fracking operation. On the following Saturday morning, people interviewed by the local press described wardrobe doors being flung open, a police station building shaking, motorbikes falling over and traffic lights suddenly not working. Pedestrians reported that the Lytham Road Bridge in Blackpool had cracked (Fig. 1.13), even though other local residents maintained that the cracks had been in existence since the 1970s and had moss growing inside them.

Earthquake seismologists from the British Geological Survey got working on the data, including seismograms and maps that plotted the focus and epicentre of the earthquake. The squiggles on the graph of the 1st April earthquake and other smaller earthquakes that had occurred at the time of the fracking were similar enough to indicate that

Figure 1.13 Cracks in Lytham Road Bridge from Daily Mail online. *http://www.dailymail.co.uk/news/-article-1372216/2-2-magnitude-earthquake-rocks-Blackpool.html. Copyright Manchester Evening News Syndication.*

the odds were that the earthquakes had been a consequence of fracking. The company involved admitted there was a problem and this resulted in a close down of fracking in Britain for almost a year.

The seismologists also pointed out, though, that the energy released by the earthquakes would have been almost impossible to feel and would not have been capable of causing structural damage. But this didn't stop people worrying. Following the earthquakes, rumours spread on the internet that fracking would cause many meters of ground subsidence even to the extent that coastal areas might witness the sea invading unless higher sea walls were built. In the south of England, rumours spread that fracking would reawaken ancient British volcanoes dormant for hundreds of millions of years. Geologists and engineers said that subsidence and volcanoes are not even remotely likely to happen. But by that time in Britain – for better or worse – shale gas and fracking already had a very bad image.

BACK TO THE FACTS?

So that's why there is a big fuss about shale gas. It seems to have a unique ability to attract interest and polarise opinion. Much of the discussion is a combustible mixture of fact and conjecture, with agendas thinly disguised on the side that wants to exploit shale and those that want to stop it at all costs. In this book I'll try to get to the bottom of this argument and to the shale itself, examining its geology and its potential, if indeed it has any outside the US.

The rest of this book will mainly be devoted to answering these questions looking mainly at the geology of shale and at the strange subsurface world of shale gas. How can groundwater be contaminated, and how can earthquakes occur? I'll concentrate on what are called peer-reviewed scientific papers – the gold standard of science – so that I can steer away from material that isn't reliable. I'll try to make these studies comprehensible to the non-geologist and I hope to help you the reader to make up your own mind, or at least have a better informed opinion.

But before I go onto the next chapter which looks at the geology of shale, I'm sure you've already noticed how mixed-up the scientific units of measurement are in the science and engineering of shale gas and energy generally. Imperial and American systems vie with standard metric (SI) units. Petroleum geologists are happy using feet and inches for dimensions and 'trillion cubic feet' for volumes of gas even though most of the world has moved on. Amounts of energy are as likely to be expressed in the modern 'joules' or older 'calories' units – or even in the rather ancient (but useful) BTU (British Thermal Unit). This mixture of units is almost unavoidable in the world of oil and gas, even in scientific papers. So I'm afraid I will have to continue to use a range of units but I'll try to explain them as I go on (if they're not self-explanatory). I also provide a simple glossary and conversion table at the back.

BIBLIOGRAPHY

AEA Technology, 2012. Climate Impact of Potential Shale Gas Production in the EU. Report for European Commission DG CLIMA. AEA/R/ED57412 67.

Christopherson, S., 2011. The Economic Consequences of Marcellus Shale Gas Extraction: Key Issues. CaRDI Reports. Issue Number 14.

Considine, T.J., Watson, R., Blumsack, S., 2010. The Economic Impacts of the Pennsylvania Marcellus Shale Natural Gas Play: An Update. Report of Pennsylvania State University, College of Earth and Mineral Sciences.

Energy Information Administration (EIA), 2011. *Annual Energy Outlook*.

Energy Information Administration (EIA), 2012. *Annual Energy Outlook*.

Energy Information Administration (EIA), 2013. Technically Recoverable Shale Oil and Shale Gas Resources: An Assessment of 137 Shale Formations in 41 Countries Outside the United States.

Energy Information Administration (EIA), 2014. Annual Energy Outlook 2014, Early Release.

Howarth, R.W., Ingraffea, A., Engelder, T., 2011. Natural gas: should fracking stop? Nature 477, 271–275.

Newell, R., 2011. Presentation to the OECD. Paris: http://www.eia.gov/pressroom/presentations/newell_06212011.pdf.

Prud'homme, A., 2014. Hydrofracking: What Everyone Needs to Know. Oxford University Press.

Rao, V., 2012. Shale Gas: The Promise and the Peril. RTI International.

Resnikoff, M., Alexandrova, E., Travers, J., 2010. Radioactivity in Marcellus Shale. Report prepared for Residents for the Preservation of Lowman and Chemung (RFPLC), Radioactive Waste Management Associates.

Shale Gas Information Platform Poland update: http://www.shale-gas-information-platform.org/areas/the-debate/shale-gas-in-poland.html.

Smith, N., Turner, P., Williams, G., 2010. UK data and analysis for shale gas prospectivity. Pet. Geol. Conf. Ser. 7, 1087–1098.

Stephenson, M.H., 2013. Returning Carbon to Nature: Coal, Carbon Capture, and Storage. Elsevier.

The Economist http://www.economist.com/news/special-report/21569570-growing-number-american-companies-are-moving-their-manufacturing-back-united.

Wang, Q., Chen, X., Jha, A.N., Rogers, H., 2014. Natural gas from shale formation – the evolution, evidences and challenges of shale gas revolution in United States. Renew. Sustain. Energy Rev. 30, 1–28.

CHAPTER 2

Shale, Shale Everywhere

Contents

A few decades ago you would have been hard-pressed to find a geologist that specialised in shale. This grey or black, fine grained sedimentary rock was considered rather boring because to the unaided eye one type of shale looks much like another. But shale has now taken on a new fascination. This is not just because shale is at the heart of a new rush for gas; it's also because shale conceals many things because it's so fine grained. In it are the remains of ancient forests and seas of algae. The fine layers – that look like the pages of a book – tell stories of ancient environmental change. The organic matter – the organic mush – that makes it dark in colour is probably one of the largest stores of ancient decayed once-living material on Earth. Just some of this organic mush is responsible for the gas that's produced by fracking. In this chapter I'll look at the formation of the shale itself, how the organic mush got in there, and how we can work out how much gas shale can produce. Then I'll explain why some places have a lot of shale gas while others have less.

Keywords: Barnett; Carbon; Fracking; Marcellus; Reserve; Resource; Shale; Source rock; Unconventional.

I'll start with the disconcerting fact that a lot of geologists don't even like the word 'shale' – because it has never been strictly defined. A geologist interested in sedimentary rocks – the kind of rocks deposited at the bottom of seas, lakes or in river valleys – would probably prefer the term 'fissile mudrock'. This technical term is a bit obscure

Shale gas and fracking
http://dx.doi.org/10.1016/B978-0-12-801606-0.00002-9

but it does convey some of what shale is. It's made of very small particles – like in mud for example – and you can split it into flat pieces.

Both of these characters – the small particles and the fine layering – are inherited from the way shale forms and they strongly influence the way it generates shale gas.

HOW SHALE FORMS

If you did any geology at school it's likely that you studied the rock cycle (Fig. 2.1). This is a simple representation of the way that all rocks are connected. Igneous rocks from volcanoes are weathered and eroded and their broken up remains get transported by rivers and landslides to seas and lakes where the pieces are deposited. The layers of deposits are called sediments. These get buried deeper and deeper under more sediment that keeps arriving and then they turn into harder more compact versions of themselves called sedimentary rocks. Sometimes if they're buried very deep they turn into metamorphic rock because of deep Earth heat and pressure. Deep rocks can get melted and

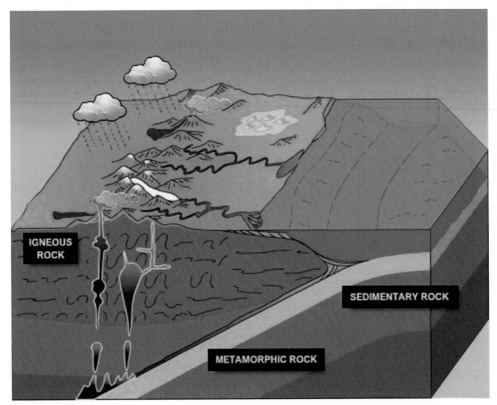

Figure 2.1 The rock cycle. *From the Geological Society of London's Website.* http://www.geolsoc.org.uk/ks3/gsl/education/resources/rockcycle/page3445.html.

included in deep magma and find their way back out onto the Earth's surface as lava, or can just be revealed by long periods of erosion. So the whole thing goes in a cycle.

To understand shale we have to see a little further into the formation of sediments. Those that are made of fragments of earlier rocks are either made of big fragments (they are coarse grained) or fine fragments (fine grained).

The fine-grained sedimentary rocks tend to be deposited far away from their sources – far from the land or mountains that they originally came from. This is a simple principle – that smaller particles can be transported by rivers further from their source than can larger particles – because they are lighter. So shale tends to be formed under deep seas and lakes a long way from the land (though there are exceptions).

There's a chemical aspect to the formation of shale also. In fact the original fragments of rock that are carried in rivers to the deep sea and lakes undergo quite a lot of degradation on the way. The minerals that make up the fragments are attacked and transformed into other minerals, usually clay minerals. These are complex but very stable, in other words they aren't likely to degrade any more. They can sit on the sea bed for a long time without changing, and then get buried under sediments that are always being fed in by rivers. So big piles of fine-grained mud collect, made up of a lot of stable clay minerals. This slowly hardens into shale. Over very long periods really huge thicknesses (several miles) of shale can accumulate.

What gives shale its fine layers? If you look closely at it, the layers are really fine laminations. These allow the shale to split, rather like the grain in wood. The laminations are very thin though. The photograph taken with a microscope (Fig. 2.2) shows a view that is less than 1 mm across so the individual layers of dark brown, light brown and light grey are only around a tenth of a millimetre thick. When you split shale with a knife or a finger nail you're spitting between these layers.

The shale has these laminations for a couple of reasons. Perhaps the most important is that the clay minerals that make up some of its bulk are naturally flat in shape or 'platey' as they are sometimes called. These tiny mineral fragments are lying flat on top of each other in the shale, piled up like minute coins. The other constituents of the shale – the remains of once-living material like plants and algae and particles of sand and silt – are also squashed in the layers, mainly because of the weight of sediments that begin to accumulate above them. It's this squashing from above that squeezes the shale into a laminated, layered texture a bit like the pages of an old book (Fig. 2.3).

As you might expect from the wide extent of the places where it is first deposited in oceans and in large lakes, shale is a very common sedimentary rock (Fig. 2.4), much more common than the coarser grained sedimentary rocks like sandstone that are deposited nearer to the sediment source. In fact shale is the most common, making up 35% of rocks at the Earth's surface. So shale isn't rare or unusual. It's ability to generate gas (or oil) has also been known for a very long time, in fact for most oil and gas geologists shale is what's known as a source rock. But more of that later.

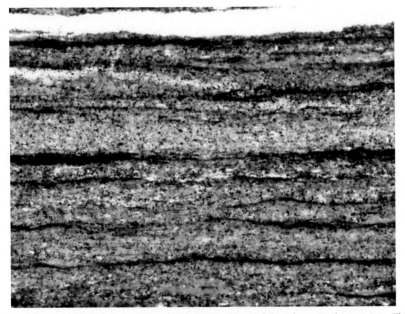

Figure 2.2 Layers or laminations in shale. The sideways span of the photo is about 1 mm. This is relatively young Palaeocene shale around 55 million years old deposited in the ancient North Sea. *Photo M. Stephenson.*

Figure 2.3 Layers of dark grey shale on a typically rainy day in the English Pennine hills. This shale outcrops in the Pennines but is present to the east and west in the deep subsurface where it is prospective for shale gas. *Photo M Stephenson.*

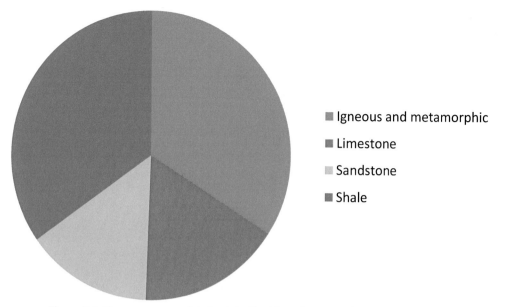

Figure 2.4 The occurrence of shale at the Earth's surface. *Data from Pettijohn (1975).*

Perhaps the most interesting part of the shale is the organic material it contains. Up to now I've been calling it mush which is clearly not a technical term but is quite appropriate. Alongside all the clay minerals and rock fragments in those ancient rivers, lakes and seas, there were a lot of small bits of once-living material – from trees, shrubs, fungi, algae, even insects and other land and sea animals. This mush gets dumped with the clay minerals and rock fragments and is completely mixed up with it. The very dark brown and black material in the microscope photograph in Fig. 2.2 is mostly organic material. Because it's very broken up and decayed, it's difficult to see precisely where the organic material has come from but careful study can help.

One of the ways we can study the organic material is to isolate it by dissolving the clay minerals, silt and sand particles from the rock so that only the organic material remains. This can be done by treating shale samples with hydrofluoric acid. After a few weeks all the rock apart from the organic material is dissolved. If you take the organic material that's left (it's usually about 1% or 2% of the weight of the rock) and spread it onto a microscope slide, it looks as shown in the microscope photo in Fig. 2.5. This picture which spans about half a millimetre, shows some of the particles that make up the organic matter.

The particles are very small and not immediately recognisable. However with careful observation it's possible to see that the elongated black particles are tiny fragments of ancient wood – in this case around 330 million years old! A scanning electron microscope can show some of the ancient structures of the wood and even help to identify the kinds of tree that produced the wood. The circular yellow-brown objects (like the one

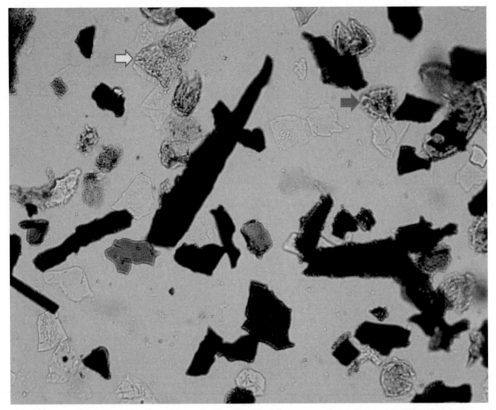

Figure 2.5 Organic matter isolated from shale of Carboniferous age from the north of England. The red arrow indicates a tree spore; the yellow arrow is amorphous organic matter. *Photo M. Stephenson.*

with the red arrow) are tree spores, perhaps spores from the same tree that produced the wood. These are less than a twentieth of a millimetre in diameter. There are other spores in the photograph, and in fact shale samples of this type and age from the north of England often contains millions of spores.

Other fragments in the photograph are harder to identify. The yellow arrow shows a fragment which has no particular shape or internal structure which helps to identify it. It would be known as amorphous organic matter (AOM) to geologists. It could be degraded wood that has rotted so much that it no longer has recognisable structures, or it could be the remains of algae or bacteria that lived in the water, or perhaps degraded marine plankton. Some research has shown that samples with a lot of AOM often contain the highest amounts of organic material and the best organic material for generating gas. So the organic matter can be mixed – some from the sea in the immediate surroundings of the sediment, some from the distant land (Fig. 2.6). This mixture of organic matter types is important to gas generation as we'll see later.

We know something of what lived in the oceans of the past because some of their teeming life was recorded in the form of fossils. In the Carboniferous shale of northern

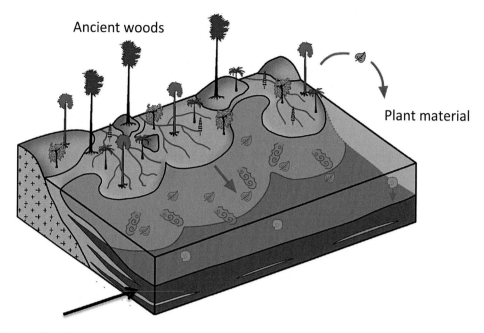

Figure 2.6 Organic material collects with the mud that eventually hardens into shale. *From Könitzer (2013).*

England for example, which is about 330 million years old, there are fragile spiral patterns that are the remains of molluscs called goniatites. They look a lot like modern *Nautilus* but are often squashed flat in the shale (the specimen in Fig. 2.7 has been lucky not to have been squashed). There are also fossil bivalve shells (like modern sea shells) and similar two-valved fossil shells known as brachiopods. Occasionally if you're lucky you might find the faint impression of ancient leaves or plant stems in shale.

But what's known as the fossil record (the collections of fossils we can see in the rocks) is notoriously incomplete. Some living things were just too soft or fragile to be preserved as fossils; some were probably too small. Perhaps the biggest question that hangs over the fossil record of shale is what produced the AOM. We'd like to know this because – as I said before – AOM appears to be the stuff that generates the most gas in shale.

If you were to take a piece of limestone or sandstone and treat it with hydrofluoric acid – to extract the organic matter – you'd find much less organic material. The proportion of organic matter in these sedimentary rocks is a minute fraction of that in shale because most of it has rotted away or been consumed in the sedimentary environment long before the sediment became rock. So why doesn't all the organic matter in shale also just rot away? The answer is partly due to the environment where shale collects. As I said earlier, the mud that becomes shale usually collects far from land in deep water. In these deep environments there's often not much oxygen and so oxidation of the organic matter doesn't happen – nor

Figure 2.7 *Gastrioceras*, a Carboniferous goniatite often found in shale. Width approximately 4 cm. *BGS Photo P549565. Photograph P549565 reproduced by permission of the British Geological Survey © NERC 2014. All rights reserved.*

does much bacterial degradation. There are often fewer animals to consume or degrade the organic matter. Environments like this are known as anoxic, and anoxia is quite common on muddy ocean bottoms where oxygenated water doesn't flow. Below a few centimetres from the surface, interstitial water (water between sediment grains) is often completely oxygen-free. Parts of the world's modern oceans are anoxic. Enclosed seas are often very inclined to be anoxic like the Black and Caspian seas. In the Black Sea, geologists have found thin light and dark layers being deposited exactly like in ancient shale and they seem to indicate seasons – a layer for each summer and winter. These layers – after many thousands or millions of years – have a good chance of turning into shale and possibly of generating gas.

How much organic matter has been buried in sedimentary rocks like shale? The physicist Philip Abelson said that '…organic matter equivalent in quantity to the weight of the earth has been created by living creatures since life originated on this planet'. He was referring to the long history of life on Earth and the mass that it had created. How much of it went into rocks? Shale is known to contain amounts of organic carbon (known as TOC or total organic carbon in the oil and gas business) of 1–10% by mass,

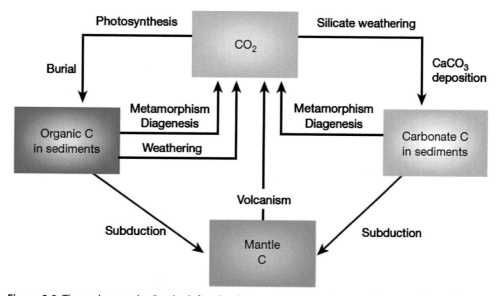

Figure 2.8 The carbon cycle. On the left side, photosynthesis creates organic matter that collects in sediments (carbon burial or capture). This carbon can remain out of the atmosphere for a very long period in 'source rock' locked away in rocks for millions of years. But eventually the carbon can be returned through weathering of exposed coal or source rock. It can also – through subduction at plate tectonic boundaries – become part of the mantle. The mantle might return it by volcanoes. *From Berner (2003).*

sometimes more. Earlier I mentioned that shale is the most common sedimentary rock by far. Because it contains much more organic carbon than other rocks (including metamorphic and igneous rocks), it's a very significant store of organic carbon on earth. I use the word store in the sense that the carbon is amongst the clay minerals and layers in deep shale all over the world, but it still can come up again. As we saw in the rock cycle, sedimentary rocks can be buried but can also be revealed again by erosion. In big earth events like mountain building, deeply buried shale appears again at the surface.

We can consider the organic carbon part of the rock cycle as an aspect of the carbon cycle which describes how carbon passes from the atmosphere to the biosphere and – in the long term – into the rocks (the *geosphere*). In an article in the journal *Nature* in 2003, the geologist Robert Berner showed how intimately related are the carbon cycle and fossil fuels (Fig. 2.8). The carbon cycle creates coal and what Berner called 'source rock'.

Berner wrote the article before the shale gas revolution in the United States and so didn't include shale gas as a fossil fuel, but 'source rock' in the sense that Berner used it is pretty much the same as shale. In fact most petroleum geologists before 2003 would probably have referred to shale as a source rock.

Let me explain. A source rock is a rock that provides the organic matter that generates oil and gas. The organic matter of source rocks was preserved and captured in oxygen-deficient mud in deep seas or lakes. So the source rock is essentially shale. After burial under other sediments, the heat and pressure begin to 'cook-up' the organic material so that some of it turns into gas or oil. We will look into the details of how this happens later but for now we'll concentrate on the fact that oil and gas have been created. If conditions are right then oil and gas can get out of the shale – for example, if it is cracked, fractured or faulted. The oil and gas tend to migrate upward and find their way into sandstone and limestone which soaks them up like a sponge (called a 'reservoir' rock). Structures called traps allow the petroleum industry to drill for them. This is the way that the industry likes oil and gas to be created because it's easy to get them out. The natural system that creates and stores hydrocarbons in convenient sandstone and limestone reservoirs has come to be known as 'conventional hydrocarbons' since the shale gas revolution. They call it 'conventional' because it's the system that the oil companies understand and like best.

So what's the unconventional system like? Well this starts the same way, with shale (or source rock) that has been cooked up enough to create oil and gas. But in the unconventional system, the shale has not released its oil and gas. The oil and gas is stuck in tiny pockets and droplets amongst the clay minerals and other organic matter in the shale. It's called unconventional mainly because the oil and gas is very difficult to get out. If you drill a well into shale nothing much happens. If you drill a well into a sandstone reservoir, the oil gushes without pumping if the pressure is high enough.

To understand why, we have to go back to the comparison I made earlier in the chapter: that shale particles are a bit like coins stacked flat on top of each other but on a microscopic scale. Imagine a big box – say a tea chest – filled with small coins all stacked flat on top of each other. The air spaces between the coins are very small, but in shale on a microscopic scale they are even smaller and there is a lot of organic matter jammed in between the flat particles. In shale, the flat particles are also cemented together by other minerals, so you can probably imagine that there isn't much space. It's very compact.

A typical sandstone rock is exactly what it says in the name – sand turned into stone. But it's completely different to shale. Like in shale, the sandstone particles aren't loose but are bound together usually by some other mineral that came along later and cemented the sand to make it solid. Using the tea chest comparison then, a sandstone reservoir would be like a filling of soccer balls or basket balls representing the sand grains. The air spaces between the soccer balls would be quite large and it isn't hard to see that the spaces would be well connected (Fig. 2.9). There might even be an equal amount of air and 'soccer ball' in the box. In sandstone the equivalent spaces are known as pores, and in this case the pores are 50% of the volume of the box. Geologists would say that the 'porosity' is 50%. A fluid would easily be able to pass through it, so we would say that the 'permeability' is high.

It doesn't take much imagination to see that the porosity and permeability of shale are lower. In fact shale is known to be thousands of times less permeable than sandstone,

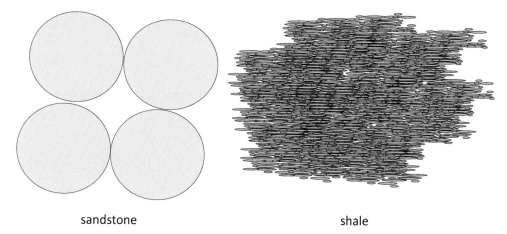

sandstone   shale

Figure 2.9 The internal structure of sandstone and shale (highly simplified) shows why permeability is so much lower in shale than sandstone. The scale is about the same for both pictures.

meaning that fluids (gas or oil) travel very slowly through shale, if at all. We can only get gas or oil out of shale naturally if the shale becomes broken or fractured by earth movements or mountain building.

So to summarise: let's follow the events through from the beginning. Mud collects on some ancient sea floor, far from the land. The mud is rich in organic matter from algae and plankton in the water column above it and from wood and other plant material that have been brought far from the land by rivers and sea currents. There's not much oxygen at the seabed and none in the sediments a few centimetres below the sea bed – so the organic material does not rot away. More mud arrives and starts to bury the other stuff. It gets deeper and deeper and begins after a very long time – maybe millions of years – to get squashed and cooked to a temperature that makes the organic material produce gas or oil. At this point if the shale is fractured by natural earth movements, the gas and oil can escape and find their way upward through their natural buoyancy to a point where they can't get any further up – where they get stuck under an impermeable layer. They may be in a porous and permeable reservoir rock like sandstone. If someone drills a hole into the reservoir rock the deep pressure of the weight of rock above will squeeze the oil or gas out into the hole. The hole will gush oil and gas. This is a conventional oil and gas system.

For unconventional oil and shale gas we have to go back to the oil and gas just being formed. If no natural fracturing happens, then the oil and gas gets stuck in the shale (in the source rock) and won't go anywhere. The low permeability imprisons the oil and gas. This is an unconventional system. To get the oil and gas out you would have to artificially fracture the shale – or frack it. This is the subject of the next chapter, but let's focus a bit more on how we work out the quantities of underground shale gas.

WORKING OUT HOW MUCH GAS THERE IS

There are two main ways to work out amounts of shale gas. Because both methods try to measure something that is deeply buried, and something that we can't see, they are approximate, so estimates of amounts of shale gas necessarily come with large error bars. The first method, which is used in some parts of the United States, uses statistics to extrapolate from well-known figures of gas yields in areas that have already been extensively drilled. It doesn't take much account of geology or even the amount of shale, but uses production figures per square kilometre to generalise to larger undrilled areas. This can be a very accurate method in areas where there is an established drilling industry.

In an area which hasn't been drilled for shale gas (which is virtually everywhere outside the US), a very different method is used. This is more fundamental, taking into account geology and the shale itself. At the heart of the method is a calculation of the underground volume of the shale and then an assumption as to its likely yield per unit volume. The method has its own uncertainties – not least the assumption about yield that has to be made – but is very instructive for those trying to understand how shale gas forms. So I'd like to talk you through a recent estimate for shale gas that was made for the north of England by the British Geological Survey (BGS). This was an important estimate because it was the first large-scale detailed look at the geological potential for shale gas in Britain. The work began by defining the area to be studied.

We know quite accurately where the shale occurs at the surface in Britain but a major effort of research was needed to understand accurately its subsurface extent. It's been known for a very long time that shale underlies much of the north of England. In fact the shale is a relic of a rather interesting period in Britain's history when it was situated close to the equator and consisted of an archipelago of islands, bays and deep-water straits, a bit like the present-day Malaysian peninsular, Sumatra and the Malacca Straits. This was the Carboniferous period and it lasted between about 360 million and 300 million years ago.

The reconstruction you can see in Fig. 2.10 shows the arrangement of the continents around 340 million years ago. The outline of present-day continents and islands was not to emerge for many millions of years, so you can't see the familiar shape of the British Isles. However, the yellow square indicates the local geography of the time – what eventually became Britain. It's interesting because this was the beginning of a period of mountain building, where huge plate tectonic forces were facing up along a line running roughly east west across the yellow square.

The south part of the yellow square became part of a huge mountain range known as the Variscan mountains by about 300 million years ago. Nowadays the remains of this mountain building can be seen in the worn-down metamorphic rock of the southwest of England and northern France, the Appalachian Mountains of northeastern United States and in the Hartz Mountains and Black Forest in Germany. What's left is only the remains of those mountains which were similar in area and height to the modern Alps.

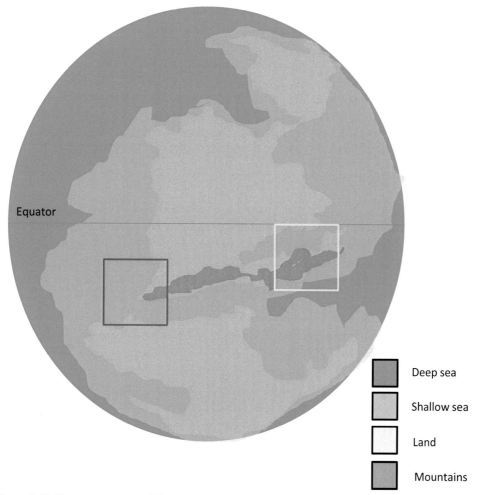

Figure 2.10 The arrangement of the continents around 340 million years ago when lots of mud was being deposited in seas and embayments at the ancient equator. The outline of present day continents and islands was not to emerge for many millions of years so you can't see the British Isles; however the yellow square indicates the geography of the time. The red square shows the ancient geography of Texas and part of the American mid-west where the mud that became the Barnett shale was being deposited. Simplified from http://cpgeosystems.com/paleomaps.html.

North of all this intense activity – in the upper half of the yellow square in Fig. 2.10 – the effects of mountain building were weaker mainly consisting of large-scale geological subsidence creating deep basins and seas. It was these deep basins that began to fill with mud. The mud came from land areas choked with equatorial vegetation and drained by huge tropical rivers. All the while the mud was collecting, the basins continued to subside, faults were formed and ancient faults were reactivated. It seems like the infilling of muddy sediment was enough to keep pace with the subsidence, so the mud reached enormous thicknesses. In the 340 million years between then and now, there was plenty of time for the mud to get squashed and harden into shale. All of this was happening in what is now the north of England. The thick shale there made the choice of an area to assess for shale gas potential quite easy. It's shown in Fig. 2.11, enclosed by the purple line.

This area includes large cities such as Manchester, Sheffield and Leeds, but also shale outcrops in the Pennine hills. The outcrops are shown in yellow. Geologists were able to visit these outcrops to gather data. This was very useful because as we'll see, most of the area within the purple line contains shale but deep below the surface so geologists can't see it directly. There are two ways to study this deep shale – through deep boreholes and seismic sections.

On the map (Fig. 2.12) you can see large black dots which show the location of 64 boreholes that go deep enough to cut through the shale either completely or partially. Usually boreholes like this are monitored and measured after they've been drilled by lowering a detection device down the hole. This device collects data about the physical properties of the rocks allowing a geologist to determine the top and bottom of the shale layer. Many boreholes are also 'cored'. This means that some of the rock that was drilled through is specially preserved inside the cylinder of the drill, behind the drill bit – like the lead in a pencil. The cylinder of rock that is brought from deep underground is known as a 'core' and is very useful to geologists because it's the only way they can see deep underground rocks directly.

Seismic sections are built up from analysis of sound waves that are transmitted into the earth by special vibrating equipment. They produce images of the boundaries between rock layers up to several kilometres below the surface, a bit like the way a ship's sonar can detect the depth of the seabed. A seismic section sometimes looks a bit like the inside of a layered cake (Fig. 2.13). A skilled geologist using information from boreholes can determine the top and base of the shale layer on the section, particularly if it passes through a location where a borehole was drilled. In this project in the north of England, 15,000 miles of seismic sections were available for study so the geologists were able to build up an accurate picture of the thickness of the buried shale and its depth.

The section in Fig. 2.13 shows the top and bottom of the Carboniferous shale layer on a 20 km long cross section in northern Lancashire. By bringing all this information together, the geologists were able to construct a 3D computer model. The model is not unlike the sort of 3D representations that auto or civil engineers use in their designs for cars or buildings, it's just that the object that is portrayed – a very irregularly shaped mass of shale – is much much larger!

It's not possible to show a 3D model on a two-dimensional page, but I can show you some images from the model. The pictures in Fig. 2.14 show the depth to the top of the Carboniferous shale and the thickness of the shale.

Greater thickness and depth is shown by the blue colour, and thinness and shallowness by the warmer orange and red colours. What you can see straight away is that shale depth is very variable. It's deep in Cheshire, Cleveland, Humberside and Bowland (near Blackpool).

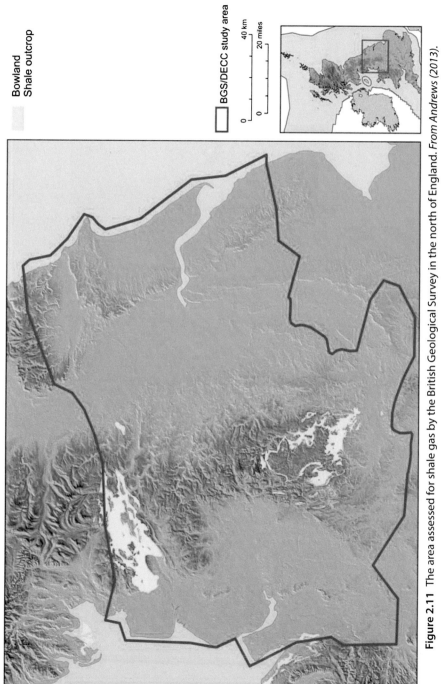

Figure 2.11 The area assessed for shale gas by the British Geological Survey in the north of England. *From Andrews (2013).*

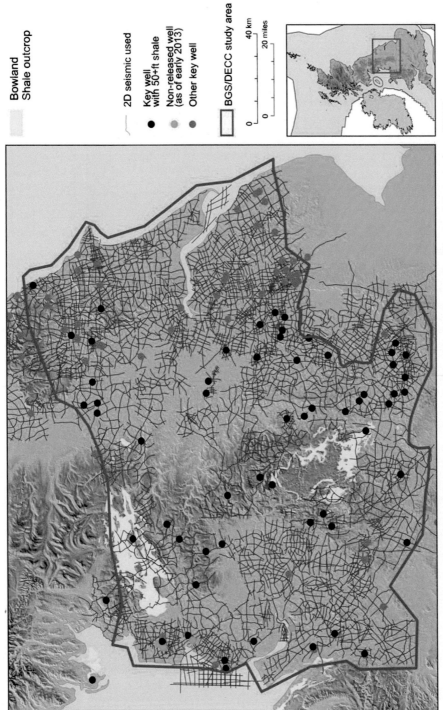

Figure 2.12 Boreholes (black dots) and seismic (blue lines) in the north of England. *From Andrews (2013).*

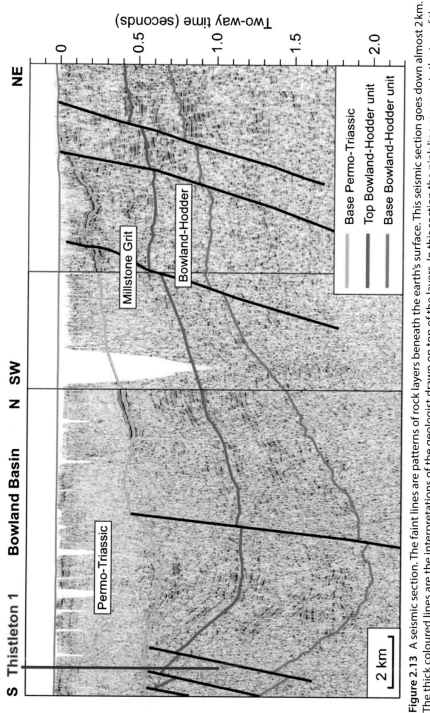

Figure 2.13 A seismic section. The faint lines are patterns of rock layers beneath the earth's surface. This seismic section goes down almost 2 km. The thick coloured lines are the interpretations of the geologist drawn on top of the layers. In this section the pink line represents the top of the main Carboniferous shale layer and the blue the base. *From Andrews (2013).*

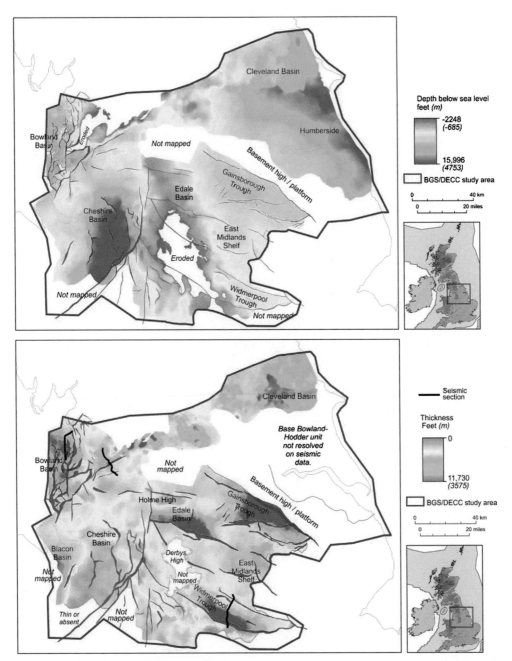

Figure 2.14 Outputs from the 3D geological model. Top: depth to the top of the Carboniferous shale. Bottom: the thickness of the shale. *From Andrews (2013).*

What surprised the geologists though is the thickness, which is really extraordinary. It's very thick in the Bowland Basin (near Blackpool), the Edale Basin (near Sheffield), the Gainsborough Trough (near Lincoln) and the Widmerpool Trough (near Nottingham). In places, as you can see from the diagram, the thickness is more than 3500 m or 3.5 km! It's greatest close to the red irregular lines in the map which are faults – geological cracks or fractures along which vertical displacement has occurred. The geologists think that the faults were moving while the mud was being deposited, lowering the sea bed slowly so that there was even more space for more mud to collect. That's probably why the shale is so thick in these areas.

There's something I didn't tell you earlier. In my discussion of the 'cooking up' of shale, I rather skirted around the details. Now it's time to return to this. In fact the shale that's been identified in the model needs a further analysis before a calculation for gas can be made. The very important point that I've ignored so far is that some of the volume or mass of underground shale may be quite unsuitable for shale gas. Some of it may not have been cooked up enough – or might have been cooked too much. Just like you can under- or overcook a cake, so you can do the same to shale.

Where does the heat come from? As I hinted before, the heat comes from burial, partly from the heat that compression produces, but also because it gets warmer the deeper down you go because you're getting closer to the Earth's core heat. You may have heard that the air temperature in deep mines can be so great that they have to be refrigerated for the miners to be able to work. At depths of several kilometres rock heat can be intense. In fact geologists know that away from volcanoes and earthquake zones, the temperature increases at the rate of 25 °C for every 1 km of increasing depth. We assume that a similar rate of temperature increase applied in the Carboniferous period but we also have ways of checking how hot shale got by looking at subtle characteristics like the shininess of the tiny wood fragments the shale contains (a property known as vitrinite reflectance). The temperature range needed to generate oil and gas is 60–120 °C, with gas being formed at the high end of this range, so this means that there is an upper and lower limit to the depth at which gas or oil might have been generated in the shale. Too shallow and the shale never got hot enough; too deep and it probably got too hot. For the northern England project, the geologists worked hard to establish these depths and this meant that quite a lot of shale didn't pass the test – in other words, it was either 'immature' or 'overmature'. The shale that passed is known as 'mature'.

After all the calculations, then, the work showed that there are about 10,000 cubic kilometres of mature shale under the north of England. A really huge amount. Using an assumption from study of American shale of a similar age this volume of rock was converted into a figure for shale *gas*: 1300 trillion cubic feet (tcf). Another huge figure.

Here's what some of that mature and immature shale looks like. Fig. 2.15 is a cross section – a diagram that geologists like to use to illustrate what they think the subsurface looks like. This cross section runs through the east of the area studied by the BGS from

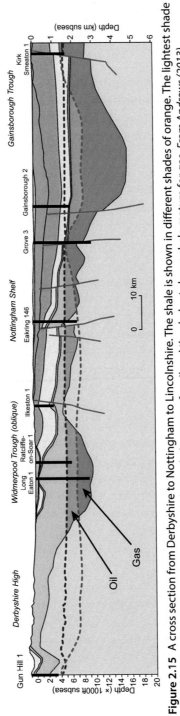

Figure 2.15 A cross section from Derbyshire to Nottingham to Lincolnshire. The shale is shown in different shades of orange. The lightest shade shows immature shale, the medium shade shows shale mature for oil and the darkest shows shale mature for gas. *From Andrews (2013).*

Derbyshire where the shale is at the surface or shallow, to south of Nottingham where it's deep, into Lincolnshire where it's even deeper.

The first thing to notice is that the shale (the three different shades of orange) varies a lot in depth. In the east in Derbyshire, the shale is at the surface. In the west in Gainsborough (Lincolnshire), it's mostly more than 2 km deep. It's also very variable in thickness. In the Gainsborough and Widmerpool troughs, it's more than 2 km thick. The other important point is that depth at which the shale is buried affects how mature it is. Most of the shale in Derbyshire is immature because it has not been buried deeply enough to generate oil or gas. The deeper shale under Gainsborough and Widmerpool is mature – the shallowest for oil, the deepest for gas. This is because oil is formed at a lower temperature than gas. So the north of England could contain 'shale oil' as well as shale gas. This was another surprise for the geologists.

The mature shale was divided into layers – an upper and lower layer – and then plotted on a map showing the main towns and roads (Fig. 2.16). The map shows that in the northeast and the west, both the upper and lower layers are present, one on top of the other. Other areas have only one layer of mature shale underneath them.

The thing that most people saw on this map when it was shown in newspapers and on television was that coloured patches coincide with populated areas like Liverpool, Manchester and Sheffield, and with towns and villages – and also with roads and railways and agricultural land. In many of these places, it's very difficult to imagine that shale gas could ever be drilled. This highlights an important point that I haven't covered yet – that just because shale gas is present doesn't mean that you can always get it out. This is at the heart of the difference between a resource and a reserve.

RESOURCE AND RESERVE

The map (Fig. 2.16) shows that shale gas may sometimes be inaccessible for reasons other than geology, for example a dense population at the surface. Another simple thing that might keep shale gas in the ground is a low gas price – one that is too low to justify the expense of getting the gas out. Policy can act in the same way, for example, if a government decides that shale gas is not right for electricity generation because it's a fossil fuel, it might act to discourage the use of gas in power stations or the drilling for shale gas, using taxes for example. The figure of 10,000 cubic kilometres of shale mature for gas and 1300 trillion cubic feet (tcf) of shale gas are really very theoretical and academic. Though I hate to say it as a geologist – they don't mean much!

In oil and gas terminology, these big theoretical numbers are known as 'resource' figures. A gas resource figure (sometimes also called a 'Gas-in-Place' figure) is the total amount that in *geological* terms constitutes material of potential value. So the 1300 tcf of shale gas is a resource figure. The idea of 'reserve' is very different. Reserve is the

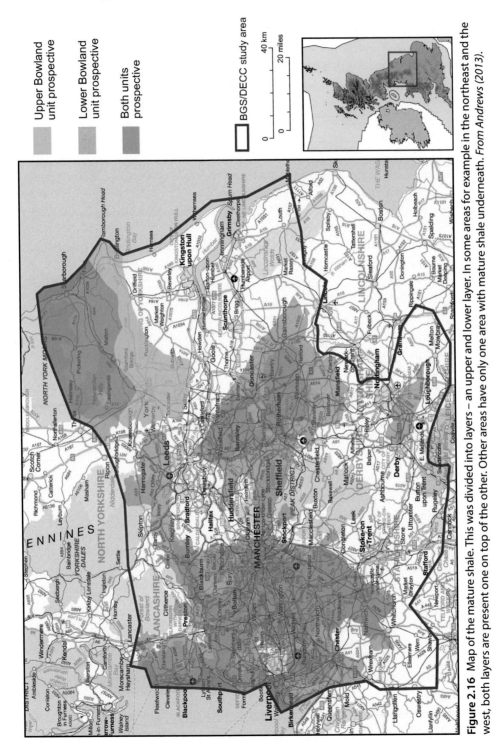

Figure 2.16 Map of the mature shale. This was divided into layers – an upper and lower layer. In some areas for example in the northeast and the west, both layers are present one on top of the other. Other areas have only one area with mature shale underneath. *From Andrews (2013).*

proportion of the resource that is economical to produce, within environmental and social limits. Another concept, the 'recovery factor' connects the two in that it's the percentage of the resources that can become reserves. The conversion of resources to reserves is the main business of technologists and scientists working in oil, and gas and minerals companies the world over. That's what geologists and engineers working for companies earn their money for.

The most common question that BGS got from the public, policy makers and industry following the publication of the north of England assessment was: what is the reserve and what is the recovery factor? Can Britain expect to extract 50% of the resource, or 10% or 1%? This sounds like an easy question to answer but it's really difficult. I'll explain why with the aid of a diagram.

Fig. 2.17 is the sort of flow chart that an oil and gas professional might have on a wall in the office. It shows the process of decision making that a company might go through to decide whether to invest in drilling wells. It's a bit like a tick list.

I'll start at the top and use the north of England example that I've just gone through. The first question is: *do I have geological understanding of the region? Do I know the resource size?*

The answer to both those questions is yes. The resource size is known. The next question: *is the resource too small or uncertain* relates to whether you can move to the next line down in the tick list. I would say that the answer is that the resource size is big enough to proceed. So we go down to the next level. Here the questions are about constraints. Are there physical, environmental or social constraints? Is the resource accessible? You could argue that this is where we are today in Britain. The companies know that there is a big resource but they don't really know whether to go to the next level. No doubt the fact that there's a dense population has an influence on them. Beyond this level are others: is there a regulatory framework in place, are there enough drilling rigs, are there gas pipelines and is there anyone to sell the gas to?

For a company to progress down the levels requires a lot of thought and effort. When and if wells start to be drilled and gas starts to be pumped, we'll be closer to understanding how much of the resource can be converted to reserve, and what the recovery factor is. But at the moment in Britain we simply don't know.

Of course in North America – where we're going next – these stages have already been gone through. In areas where there's a lot of drilling, recovery factors are known pretty thoroughly. In these places no one really cares what the resource is – only the reserve. The reserves are really important and are very precisely known. In fact the reserve associated with a particular company is essentially the company's main asset and is very closely connected to its value for obvious reasons. It's like money in the bank.

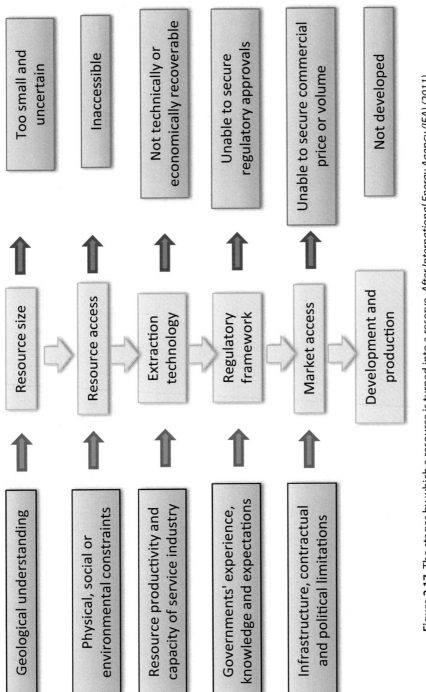

Figure 2.17 The stages by which a resource is turned into a reserve. *After International Energy Agency (IEA) (2011).*

NORTH AMERICA

We saw in the last section that the most important things needed for shale gas are shale with plenty of organic matter, layers of the right thickness and the right pattern of maturity. We can follow these principles to see how shale gas is distributed in the main North American shales, the Barnett, Marcellus, Fayetteville, Haynesville, Eagle Ford, Muskwa and Montney (Fig. 2.18).

Barnett Shale

As it happens, virtually the same conditions that produced shale in Britain in the Carboniferous period also existed in the US. In fact the Barnett shale – which you'll remember from the first chapter was the first to be exploited on an industrial scale – is of a similar age to the Carboniferous shale of the north of England but a few thousand miles away to the west (see Fig. 2.10). One difference is that the Barnett shale was formed in more open marine conditions with fewer land areas nearby. The area over which the shale was formed was also much larger: tens of times larger in fact. Interestingly the Barnett is not so thick though, perhaps because it's less affected by the faults that encouraged great thicknesses in the north of England.

The Barnett shale is deep underneath an area of 28,000 square miles in north–central Texas, as well as the city of Fort Worth. Many other rock layers are squeezed between it and the ground surface. The shale only comes to the surface as small outcrops in the south in the Llano area where it's well known for having very high levels of organic matter – up to 13% by weight. So it's an extraordinarily rich source rock. These shale outcrops have never been buried deeply and so the shale is immature in this area. A quick look at the map (Fig. 2.19) shows that the Barnett shale descends deep into the subsurface north from Llano. The red contour lines are like topographic contour lines in that they illustrate the shape and inclination of a surface or a plane. In this case, the plane is underground and it's the base of the Barnett shale.

The grey area is the current main production area of the Barnett, known as the Newark East gas field. Two reasons for the positioning of the field are that the shale is mature for gas in that area mainly because of the great depth of the shale, and that the thickness of the Barnett shale increases from 500 to 700 feet in that area. So the Newark East gas field is in the right place – where the shale is mature and thick – and for most shales in the US the same principles apply. The Newark East field is currently the largest gas field (unconventional or conventional) in Texas covering 500 square miles, with over 2400 producing wells and 2.7 tcf of proven reserves (not resources!).

Marcellus Shale

The Marcellus shale is different from the Barnett in being quite a lot older. The mud that formed the shale was deposited on an ancient sea bed 380 million years ago not far from

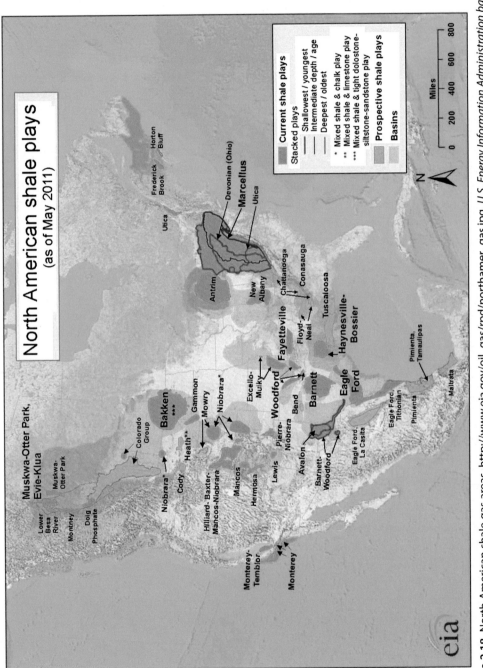

Figure 2.18 North American shale gas areas. http://www.eia.gov/oil_gas/rpd/northamer_gas.jpg. *U.S. Energy Information Administration based on data from various published studies. Canada and Mexico plays from ARI. Updated: May 9, 2011.*

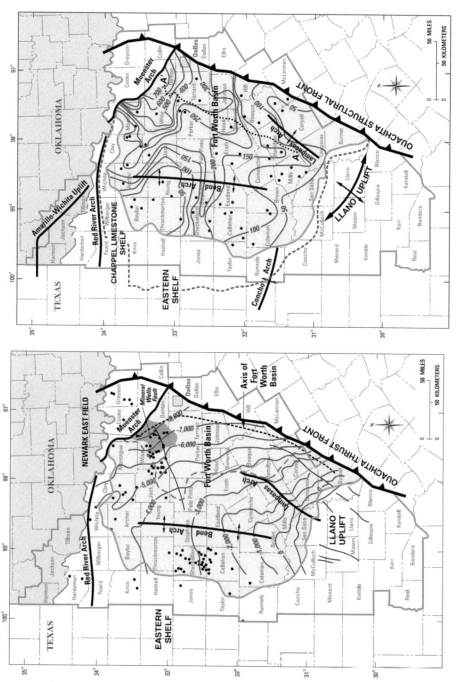

Figure 2.19 Maps of the Barnett shale. The left map shows the main producing area of the Barnett shale which is known as the Newark East field (dark grey patch). The red contour lines (interval 1000 feet) show the depth to the base of the Barnett shale. The shale gets deeper north and eastward. The map on the right has red contour lines that show the thickness of the Barnett shale (intervals are 50 and 100 ft). The shale is less than 250 feet thick over most of the area but in the northeast around the Newark East field it reaches 500–700 feet thick. *After Bruner and Smosna (2011).*

the long-vanished Acadian Mountains, out past an ancient river delta known as the Catskill delta. The shale extends for almost 600 miles through Ontario, New York, Pennsylvania, Ohio, West Virginia, Maryland and Virginia, over 75,000 square miles.

The secret to this shale's productivity again is its high organic matter content (up to 11%), its thickness and the pattern of maturity that it inherited from deep heat in the crust but also the proximity of an active chain of mountains. The Acadian mountains built up in what is now south eastern Pennsylvania, Maryland and Virginia and their remnants – the Appalachian Mountains – form the edge of the area that is underlain by the Marcellus shale.

These rather complicated maps (Fig. 2.20) reveal the key to the most productive areas. On the left there is a 'contour map' that shows the thickness of the Marcellus shale through Pennsylvania and New York State. The blue green colour indicates the thinnest shale and the red the thickest shale. In most cases the drillers will want to go for the thickest shale to maximise their yield and in northeast Pennsylvania it reaches its maximum thickness of more than 250 feet. On the right is another contour map showing the maturity of the shale (how much it has been cooked up). In this map, the reds and greens are the best colours and the blue colour to the west is mostly immature.

These two features alone are enough to explain the best shale to drill, which runs in a belt from the northeast in New York, through northeast Pennsylvania into southwest Pennsylvania (Fig. 2.21). But it doesn't include the far west of Pennsylvania and West Virginia where the shale is either immature or thin (though there are exceptions). The core area – where the formation exceeds 50 feet thick and has the right maturity – covers a huge 50,000 square miles making it the largest potential shale region in North America.

Eagle Ford Shale

The Eagle Ford shale, named after the small town of Eagle Ford a few miles west of Dallas, Texas, is much younger than either the Marcellus or the Barnett coming in at a mere 90 million years old. It's underneath much of south Texas into Mexico but the best shale trends across Texas from the Mexican border up into East Texas, in an area roughly 50 miles wide and 400 miles long. Here the shale has an average thickness of 250 feet. The subsurface structure of the Eagle Ford shale is relatively simple in that it's part of a set of rock layers that is inclined down towards the southeast like a tilted stack of books. The northwest end – where the shale outcrops – makes a long curving stripe between the Mexican border, San Antonio and Austin. From this point, the shale dives deep below the surface getter deeper to the southeast.

The pattern of oil, mixed oil and gas, and just gas ('dry gas') illustrates nicely the principle that I mentioned in earlier discussions of the Barnett and Marcellus – that generally as you go deeper maturity increases (Fig. 2.22). The coloured bands showing Eagle Ford oil and gas

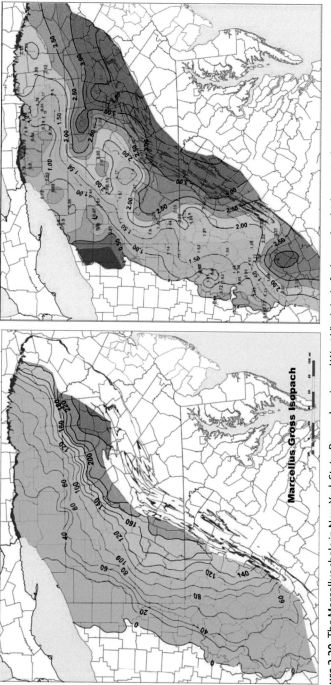

Figure 2.20 The Marcellus shale in New York State, Pennsylvania and West Virginia. Left shows the thickness (in feet) of the Marcellus shale. Green and blue-green is thin; orange and red is thick. The thickest part is in northeastern Pennsylvania. Right shows the maturity of the shale. Blue and blue-green is mainly immature, red is mature, so the shale is mature in a belt running diagonally across Pennsylvania and into West Virginia. *From Gregory Wrightstone (2009). Adapted from extended abstract prepared for oral presentation at AAPG Annual Convention, Denver, Colorado, June 7–10, 2009.*

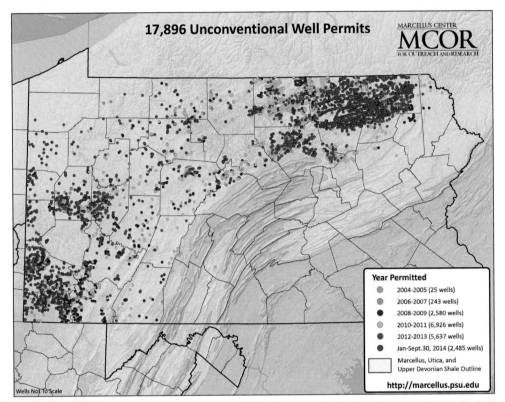

Figure 2.21 Drilling permits issued for Pennsylvania from 2004 to September 2014 showing the belt running from northeast of the state to southwest where the best shale is present. Most of the permits will be used to drill the Marcellus shale. *From the Marcellus Centre for Outreach and Research, Penn State,* www.marcellus.psu.edu.

appear about 50 miles southeast of the outcrop. This is because in the area between, the shale isn't deep enough to be mature. Because oil is the first hydrocarbon to be generated on the route to full maturity, oil appears in the shallowest part of the productive area. Pure ('dry') gas doesn't appear until further southeast where the shale is even deeper.

Though the Eagle Ford Shale is a relative newcomer in American shale, it seems to have the potential to produce more gas and oil than other shales. This maybe because fracking of Eagle Ford shale is more efficient because it contains minerals that make the shale more brittle and 'frackable'. We'll discuss this in more detail in the next chapter.

Haynesville Shale

The 150 million-year-old Haynesville shale underlies an area of about 9000 square miles of east Texas and west Louisiana averaging about 200–300 feet thick in the main productive area. The shale is very deep in the area occurring 10,500 to 13,000 feet

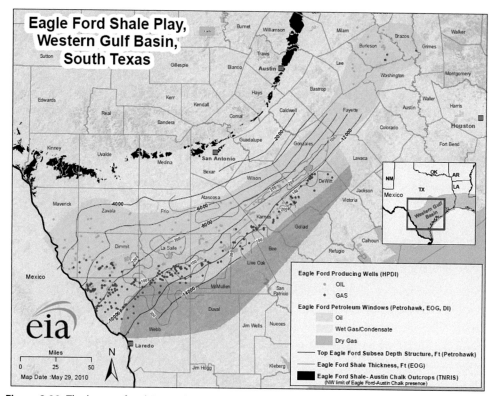

Figure 2.22 The layers of rock in south Texas are inclined downward to the southeast so that maturity in the Eagle Ford shale increases in that direction. This explains the pattern of yields of oil and gas. *From the EIA website:* http://www.eia.gov/…'

below the surface, and produces mainly gas. This makes it quite expensive to drill. The Haynesville shale was one of the fastest growing shale gas businesses in the US until about 2012 when gas prices fell and it became more difficult to operate profitably. Many companies shifted to more profitable shales where they could get oil as well as gas, for example the Eagle Ford shale of south Texas.

Fayetteville Shale

The Fayetteville shale is roughly the same age as the Barnett shale and underlies much of northern Arkansas and adjacent states. It produces natural gas in the central portion of the Arkoma basin in Van Buren and Cleburn counties where the shale is at a few hundred to 7000 feet below the surface. To the north in Arkansas, the Fayetteville shale outcrops in a broad arc.

Muskwa (Horn River) Shale and Montney Shale in Canada

North of the border, two shales are important: the 375 million-year-old Muskwa (Horn River) shale and the 250 million-year-old Montney shale. The Muskwa Shale was first described in outcrop on the banks of the Horn River, a tributary of the Mackenzie River, in the Northwest Territories. In eastern British Colombia, it ranges in depth from 6300 to 10,200 feet, averaging 8000 feet for the prospective area, and organic matter averages 3.5%. The maturity is high so gas is common.

The Montney shale in northwestern Alberta covers an area of approximately 1900 square miles at a depth ranging from 3000 to 9000 feet, averaging 6000 feet for the prospective area. The organically rich part of the Montney shale averages 400 feet thick and the organic matter constitutes about 3%. Its high maturity means that gas is mainly produced.

IT ALL SEEMS SO SIMPLE

We've seen that shale is a common rock that was once mud deposited in deep seas and lakes where the organic matter – the wood, spores, algae – don't degrade but become preserved and entombed. The mud gets squashed and hardens to shale and begins to heat and pressurise with burial so that eventually hundreds or thousands of feet below the surface and millions of years after it was buried, the organic matter begins to make gas or oil or both. If the shale is fractured naturally or broken by faults, this oil and gas might escape and under its own buoyancy find its way up through the rocks above into natural arches. In these 'traps' it's still straining to get further up – so if you drill it will gush from a well. If the shale remains untouched and unfractured, it's just too impermeable to allow the gas or oil to go anywhere. To get oil and gas you have to artificially fracture the shale.

This all seems so simple but it isn't at all. Hydraulic fracturing in shale is a high tech, high capital business and it's the subject of the next chapter.

BIBLIOGRAPHY

Abelson, P.H., 1957. Organic constituents of fossils. Treatise on marine ecology and paleoecology. 2 Mem. Geol. Soc. Am. 67, 87–92.

Andrews, I.J., 2013. The Carboniferous Bowland Shale Gas Study: Geology and Resource Estimation. British Geological Survey for Department of Energy and Climate Change, London, UK.

Berner, R., 2003. The long-term carbon cycle, fossil fuels and atmospheric composition. Nature 426, 323–326.

Bruner, K.R., Smosna, R., 2011. A Comparative Study of the Mississippian Barnett Shale, Fort Worth Basin, and Devonian Marcellus Shale. U.S. Department of Energy, Appalachian Basin. Report DOE/NETL-2011/1478.

Energy Information Administration (EIA), 2013. Technically Recoverable Shale Oil and Shale Gas Resources: An Assessment of 137 Shale Formations in 41 Countries Outside the United States. Energy Information Administration website: http://www.eia.gov/.

Hammes, U., Hamlin, H.S., Ewing, T.E., 2011. Geologic analysis of the Upper Jurassic Haynesville Shale in east Texas and west Louisiana. AAPG Bull. 95, 1643–1666.

International Energy Agency (IEA), 2011. Are We Entering a Golden Age of Gas? Special Report of the World Energy Outlook.

Könitzer, S.F., 2013. Primary Biological Controls on UK Lower Namurian Shale Gas Prospectivity: Understanding a Major Potential UK Unconventional Gas Resource. Department of Geology, University of Leicester, unpublished Ph.D thesis.

Pettijohn, F., 1975. Sedimentary Rocks, third ed. Harper.

Wrightstone, G., 2009. Marcellus Shale – Geologic Controls on Production. Search and Discovery Article #10206. American Association of Petroleum Geologists.

CHAPTER 3

To Frack or Not to Frack?

Contents

Geologists and engineers aren't known for their careful language. Words that sound alarming abound in the geological literature: 'supercritical' for deep gas like CO_2 when it becomes pressurised; 'overpressured' for layers of rock that have higher pressure than their depth would suggest; 'blowout' for when a well leaks gas uncontrollably. It's probably because geologists and the drilling and mining engineers that they work with don't imagine that their technical language will ever go beyond the technical literature, the companies, the research institutes and the universities where their techniques are discussed. However, fracking (hydraulic fracturing) is a phrase that has gone far and its hard sound and its similarity to an Anglo-Saxon expletive is a kernel around which the arguments coalesce. In much protest literature, the word has become synonymous with shale gas even though fracking has been used for a long time in conventional oil and gas and even in geothermal power.

Whether the technologists and engineers picked a good name or not, the technique is being used extensively across North America and is being tried in China, Europe and South America. It involves squirting fluid under very high pressure into rocks far below the surface to open up existing cracks and make new ones so that gas can be released. This sounds easy but it isn't. There are a whole host of problems relating to the depth of the rock, the properties of the fluid and to getting the gas out.

But engineers and geologists have by and large solved the technical problems and fracking has never been used so widely or so profitably. About 33,000 gas wells are drilled in the US every year and about 90% of these are fracked to get gas out. This chapter will look at how fracking works.

Keywords: Frack fluid; Fracking; Fractures; Horizontal well; Hydraulic fracturing; Proppant; Shale.

Why are cracks or fractures so important for shale gas? To understand this, we have to go back to the last chapter where I compared sandstone with shale. You might remember that sandstone is a conventional reservoir for oil or gas, which essentially means that it's easy to get the stuff out. Shale is compact and impermeable. Gas that has been formed in the shale from cooking the organic mush will tend to stay put unless something allows it to move. What allows it to move are fractures. Fractures make shale permeable.

Shale gas and fracking
http://dx.doi.org/10.1016/B978-0-12-801606-0.00003-0

Figure 3.1 Natural cracks in Marcellus shale in a stream bed in New York State. Courtesy of Dr Gary Lash.

In fact all rocks, including shale, have natural cracks or fractures. The photograph (Fig. 3.1) seems to show a nicely constructed paved path in a stream but in fact it's naturally occurring shale that outcrops in a stream. This is a similar shale to the Marcellus shale that's being drilled at a depth of thousands of feet below the surface in Pennsylvania, but at this location in rural New York, it's right at the surface. The reason that it has this rather neat blocky appearance is because of the natural fractures that it contains, and in the outcrop in the photograph the two sets of fractures are almost perfectly at right angles. So the shale looks like flagstones lain in a garden.

Cracks like these form in solid, hard rock when it's stretched or pulled apart. The spacing between the cracks depends on the strength of the rock and the thickness of the layer. People who own old houses will know these kinds of cracks well – they form above windows and doors where there is some stress in the structure. Normally the crack will open in the direction parallel to the strongest stretching or tension.

It's fairly obvious that walls of houses can be stretched, but you might ask how rocks get into this situation. Anyone who's seen a TV documentary on mountains or plate tectonics will know that geological plates and continents are colliding, squeezing up against other – and have been for hundreds of millions of years. So we know that rocks get squashed. But near large faults, for example, strong tensions can pull rocks apart as well. In less dramatic circumstances, a deeply buried rock – that's slowly exhumed by erosion of the layers above it – literally relaxes and spreads both upwards and sideways. It's a bit like the top of a cake once it breaks out of a cake tin and the surface cracks.

As I said, fractures make the shale permeable and it's through these that gas can move. So that's why fractures are important. The fractures that you see in the photograph were probably once filled with methane gas that's now long gone because it will have escaped to the atmosphere. The gas in the fractures would have come from the shale around them slowly travelling through the unfractured shale creeping a few millimetres every few thousand years. But in the same Marcellus shale deep below the surface – deep under the farms and woods of Pennsylvania – the fractures will likely still contain the gas because it's not exposed to the surface and it is confined all around by shale and more shale.

For a shale gas driller, the photograph in Fig. 3.1 is very interesting because it means that there might be fractures far below the surface and because the fractures are very well connected. In the picture, two sets of parallel fractures intersect at right angles. The driller would find this encouraging because his drill hole has only to connect with some of the natural fractures, and gas will flow to his well. It will rush through the right-angled passageways in a mad dash to get into the well. This is because of the pressure difference between the well and the rock thousands of feet down. The weight of rock above will squeeze the gas out into the well. So won't the drilling engineer be content with that – gas flowing freely from the shale into the hole?

The answer is no. In the first chapter, we saw that shale is filled with gas if it's undergone the right cooking and if it had enough organic mush to begin with. Every gap and space will have molecules of methane. Looking back at the photograph (Fig. 3.1), you can see that if the well connects with the natural fractures of the Marcellus shale (if they are the same deep down) will only get you access to the gas in the fractures. There's an awful lot locked away in the squares or cubes of rock between the fractures. If you're going to spend millions drilling a hole then you'll want some of that gas too. You might also find that fractures deep down are squeezed so tightly together because of the pressure that they don't let gas flow. Or they might be choked up with other minerals formed later.

So you need to 'open the shale up' a little by widening the existing fractures, extending them and even making new ones. You also have to drill in a direction that connects with as many natural cracks as you can. The more cracks you can make and the more cracks you can intersect, the more gas you can get. To make new cracks, the driller pumps fluid – usually water and chemicals – down into the rock layer at truly immense pressures and then props the new cracks open with sand (called proppant) so they don't close after you've reduced the pressure.

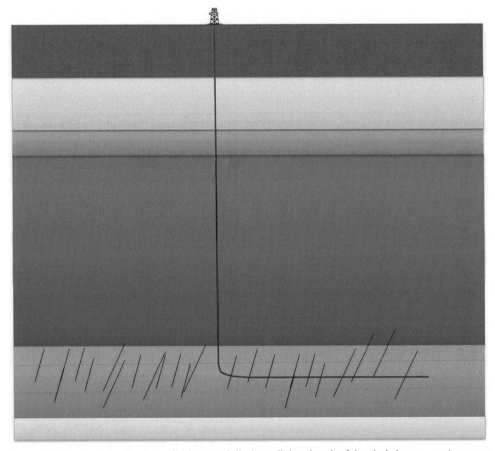

Figure 3.2 Horizontal wells – or wells that are drilled parallel to the tilt of the shale layer – can intercept many more natural cracks and more shale.

If you want to intersect as many cracks and as much shale as you can, drills need to follow layers. In Pennsylvania and West Virginia, the best Marcellus shale is about 50-feet thick under 50,000 square miles of country. Over that area the layer is often horizontal. If you drill a vertical hole through the shale you'll only intersect 50 feet of shale and 50 foot-worth of natural cracks before you reach the base of the shale layer. So it makes sense to try to follow the layer horizontally so that your well can intersect many more natural cracks and more shale. These are horizontal or lateral wells (Fig. 3.2).

FRACKING IN ACTION

In October 2013, I visited a fracking operation in Fox River, Alberta. Flying over a landscape of conifers and birch woods and wetlands, it was clear how wild and remote the area is. But 3 km below the woods is an organic rich layer – the Duvernay shale – which

Figure 3.3 A fracking site at Fox Creek, Alberta. *Photo M Stephenson.*

is 35 to 60-m thick and extends over much of northwest Alberta. The operations look insignificant from the air – football field-sized areas packed with machinery, trucks and pipes – with miles of woods between. In the Fox Creek area, 195 horizontal wells have been licensed and drilled and about 50 of those are producing gas.

On the ground these football field-sized areas known as well pads appear even busier.

Trucks deliver diesel to power the generators and the frack trucks, or bring proppant sand. In some fracking operations, water is brought by road tankers. Men in protective clothing and hard hats walk purposefully around checking valves and moving equipment (Fig. 3.3).

But this is not what assails you immediately. It's the noise. The Fox Creek operations are what are known as a 'high volume' fracks which mean that very large amounts of fluid and proppant are forced down the well. To do this you need a lot of power. The fluid is pushed down the well at pressures as high as 700 atmospheres – in other words 700 times atmospheric pressure – or around 10,000 pounds per square inch (psi) or 70,000 kPa. A line of enormous 'frack trucks' delivers the power (Fig. 3.4).

Let me explain why there's so much sound and fury. The wells in this part of Alberta drilled into the Duvernay shale are 3 km or more deep. But they don't stop there. At just about that depth, the drillers alter the force in the drill bit so that it turns from vertical to horizontal. The horizontal part of the well then follows along the shale layer for hundreds or even thousands of metres so that there's as much contact as possible with shale and fractures.

That means the length of the hole you're drilling could be 5 km or 3 miles. The internal diameter is around 10 cm. Anyone with a long hosepipe and a big garden knows that to get water down a long hosepipe takes quite a lot of pressure from the house water supply. It also takes a while for the water to appear at the far end of the hosepipe. That's because it takes quite a lot of water just to fill the pipe.

In a 3-mile-long hole you obviously need even more water, but consider the fact that you also want that water and the proppant that it contains to run into existing cracks and open new ones. To get the best yield of gas, engineers want to open up the shale by widening cracks and making new cracks for many metres either side of the well. The *volume* of this cracked rock could be many thousands of cubic metres.

How wide are the cracks? Well it depends – partly on the width of the natural cracks and partly on the width of the cracks you hope to make. For the new cracks, their width will probably be governed by the diameter of the sand grains that you're using to prop them open. This is usually a fraction of a millimetre. So the volume represented by a single crack may not be much, but the volume of all the cracks added together could be very large consuming a lot of water. In fact during a high-volume frack up to 250 litres of water can disappear down the well every second. So this is why you need a lot of water and proppant in a high-volume frack.

Engineers often mention the phrase 'frack fluid' rather than water. This is because although the fluid they pump down the well is overwhelmingly water there are chemicals added to the water apart from the solid sand proppant.

The most important chemical is a friction reducer. To understand why this is needed, think back to the long hosepipe. With a really long hosepipe, part of the reason why you have

Figure 3.4 A line of frack trucks at Fox Creek, Alberta. *Photo M Stephenson.*

to force the water through the pipe is that there is friction between the water and the walls of the hosepipe and internally within the water. The friction effect is very large in a deep well, particularly when water is being forced into a myriad of tiny cracks 3 miles down the pipe. Interestingly though, the water that's carrying the proppant also has to – at the right time – become viscous enough to carry the proppant to where it's needed. So you want the frack fluid to be non-viscous (runny) in the early parts of the frack and then viscous (not runny) later on! This is achieved by a cocktail of chemicals added at the surface.

The pie charts (Fig. 3.5) show the typical composition of frack fluid. Most of the liquid is water but the sand proppant is an important addition. On average 0.17% of the volume is made up of chemicals. These might include an inhibitor to prevent scale on the walls of the well (like the fur in your kettle), acid to help initiate fractures and biocide to kill bacteria that can produce acids that lead to corrosion.

The fracking is not as simple as it may appear because although I've described pumping of frack fluid into the well, the well itself is not a simple open hole. Soon after the drilling and before any fracking is carried out, the well has a 'casing' inserted into it, all the way to the bottom. The casing is a steel pipe that is assembled at the top of the well and inserted piece by piece. It's usually held in place with cement which bonds the outside of the pipe to the rock and seals the space between the outside of the casing and the rock. The main reasons for having a casing are to stop the hole caving in (which would be a big waste of money) and to stop anything getting from the inside of the well into the rock that it passes through. We'll see why that's important later.

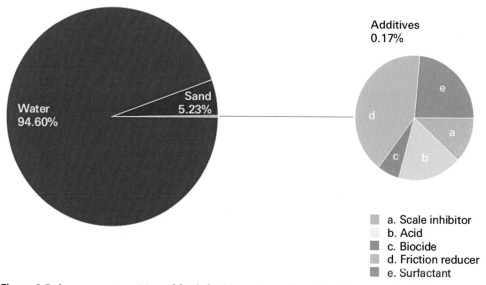

Figure 3.5 Average composition of frack fluid by volume. *From Royal Society and Royal Academy of Engineering report (2012).*

There are several kinds of casing all with different diameters for use at different depths in the hole. Rather like a telescope the casing gets narrower the further down the well you go. The next casing can be slid through the previous one reaching ever greater depths. In the diagram (Fig. 3.6) these are shown compressed into a short space though of course each segment of casing might be hundreds or thousands of metres long. The most important part in the business of fracking is the final production casing that's cemented into the shale to support the hole at great depth. It's also there to have tiny holes punched through it. These are the holes that concentrate the high-pressure frack fluid forcing it into the shale to make the cracks – and then let the gas out.

In the picturesquely named *plug and perf* method, several sets of holes are made in the production casing. The engineers start at the end of the horizontal well – what's known as the *toe* of the well for obvious reasons. The engineers want to focus the pressure on one part of the well to get the best effect – so they push perforating guns with explosive charges that blow tiny holes in the production casing right to the toe. These do their work and then this part of the well is high pressure fracked. The rest of the horizontal part of the well is sealed-off at this time by a plug inside the production casing that stops the pressurised frack fluid getting out. Once this is done, the next bit of the horizontal part is fracked moving back toward the *heel* of the well. This is a *multistage frack* and it's a way of creating the maximum number of cracks possible in the shale along the horizontal part of the well. After the frack is finished, the plugs are drilled through and frack fluid is allowed to flow back out of the well. It's usually kept in tanks at the surface for recycling on other fracks (Fig. 3.7).

As soon as the pressure is reduced after fracking when all the pumps are switched off, the pressure difference around the well is reversed. The rock is now under the highest pressure and the inside of the well (inside the production casing) is lower pressure. This is useful because the natural rock pressure is now what forces the gas out into the well – and this pressure is greater with greater depth. This is one of the reasons why fracking can't take place at shallow depths – because the rock doesn't squeeze the gas out very effectively.

The first fluid to come out is the frack fluid and this starts to flow immediately. It's known as 'flowback'. The first gas might not appear for between 2 and 20 days after fracking (Fig. 3.8).

Interestingly, not all the frack fluid injected into the shale layer is recovered. The amount of recovered fluid can be as little as 5% in the Haynesville shale to 50% in parts of the Marcellus and Barnett shales. The main reason for this appears to be that most shale is undersaturated with water (unlike most other sedimentary rocks), in other words there's very little water within the spaces between the shale particles. When the gas starts to flow out, the frack fluid enters some of the spaces and small fractures and stays there. This is one of the reasons why it isn't possible to recycle all of the flowback and why engineers have to keep topping up their water supplies.

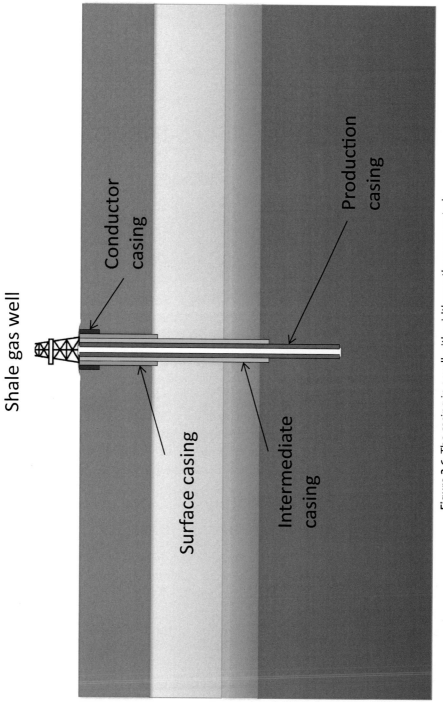

Figure 3.6 The casing in a well with width greatly exaggerated.

Figure 3.7 The tanks in the background store frack fluid for reuse. *Photo M Stephenson.*

The flowback also contains things it didn't have when it was pumped down the well, for example it may contain radioactive substances and dissolved substances that it picked up while it was being squeezed through the shale far below. This means that flowback water that you don't recycle can't be disposed of in a simple way into the environment. We'll look at this in a later chapter.

HOW DO ENGINEERS KNOW WHERE THE FRACTURES GO?

The machines and power involved in fracking make it very expensive. To make money, a lot of gas has to be extracted. So companies want to target precisely the shale that will yield the most gas. The engineers don't want to frack outside of the best shale because they're wasting the pressure that they've spent a lot of energy building up. To monitor the progress of the frack they've invented methods such as microseismic monitoring.

When the high-pressure frack fluid starts cracking the rock as you'd expect, vibrations are sent through the rock in all directions, a bit like popping sounds. Detectors can be placed in monitoring wells all around the well that's being fracked. If the vibrations

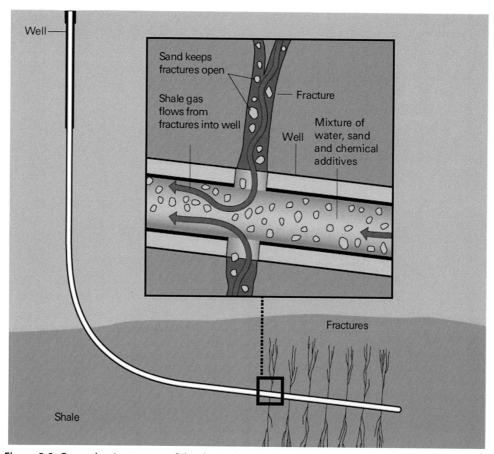

Figure 3.8 Gas makes its way out of the shale, along fractures and into the production casing. Then it rushes up the well to the surface. *From Royal Society and Royal Academy of Engineering report (2012).*

are picked up at several locations then the sources of the vibrations (the positions of the cracks that are forming) can be plotted. Such a plot is shown in Fig. 3.9 as a cloud of coloured dots. The different colours represent groups of new fractures from single fracking stages. You can see from the diagram immediately that the well has been fracked several times.

The light blue dots represent one of the early frack stages at the toe of the well. The dark blue shows a later frack at the heel. You can also see that almost all of the cracks were created within the Barnett shale which is good because that's where the gas is concentrated.

But what if a shale layer is very thick or extends over a very wide area? If a company is going to invest in fracking where will they choose first? Most companies want to maximise their chance of getting gas first time. If they can, they'd like to know exactly

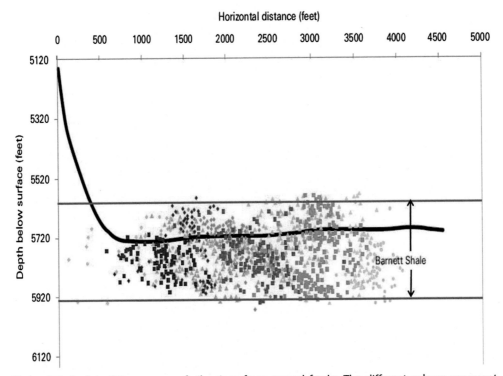

Figure 3.9 A plot of the sources of vibrations from several fracks. The different colours represent groups of new fractures from single fracking stages. The light blue dots represent one of the early frack stages at the toe of the well. The dark blue shows a later frack at the heel. *From Royal Society and Royal Academy of Engineering report (2012).*

where to drill – mainly because drilling a well is so expensive. Apart from the kinds of maps we saw in the second chapter showing the depths and thicknesses of shale, what can be done to pinpoint the right place to drill? This is where the geologists really come into their own.

SWEET SPOTS

Geologists find what are called the *sweet spots* in the shale – the places where the best results can be expected, like the raisins in a cake.

Earlier I mentioned how seismic surveys can be used to draw cross-sections of the deep underground. The principle is simple – like a marine sonar, vibrations can be used to detect the tops and bottoms of deep layers because each layer has slightly different properties and so boundaries between layers act as reflectors. But when vibrations come back, there are far more complex vibrations than just those from the

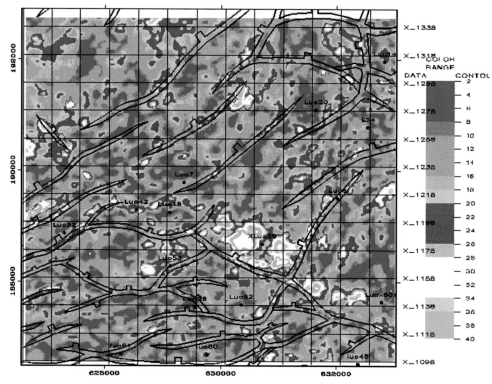

Figure 3.10 Natural fractures in a shale layer in the Yellow River area, eastern China, have been detected using sophisticated seismic processing techniques. The colours on the map represent the density of natural fractures, green being the highest density. *From Li et al. (2003).*

tops and bottoms of layers. There are also more complex vibrations that include reflections from deep fractures, particularly if those fractures are common and parallel. This means that with sophisticated mathematical processing of seismic data, it's possible to identify the distribution, orientation and density of fractures at great depths in rocks. And this is without even drilling a hole. An example is shown above (Fig. 3.10). The colours on the map represent the density of natural fractures in a shale layer in eastern China. These are fractures like those we saw in the photograph of the Marcellus shale in the river bed in Fig. 3.1, but these fractures are deep under the fields of China. The concentration and the orientation of the fractures could help the drillers choose the best place to drill and the best direction in to drill the horizontal part of the well.

As well as knowing where to drill and what direction to point the horizontal part of the well, it might be important if you have a very thick shale layer to know at what depth to start the horizontal part.

In an earlier chapter, I described how layers of mud collect one on top of another. Those layers contain the mush of organic matter. The mush is from the plants on the land around and from the algae and plankton growing in the water. Because the conditions on the land change through time and because the sea also changes its productivity and ecology, the mix of organic matter in the mud and eventually in the shale changes layer by layer. Some layers may have a better mix than others for making shale gas.

The rather complicated diagram below (Fig. 3.11) shows what I mean by sweet spots in shale layers. It's worth looking into in a bit of detail – so here goes.

The diagram shows a series of layers of shale in a core from a borehole near Derby in Britain. The shale shown is the Carboniferous shale of the north of England that was the subject of part of the second chapter. On the left is a column with symbols showing the variation in the core over 55 m. The top of the column is the top of the core (actually at the ground surface). The bottom of the column is deeper in the shale, 55 m below the surface. But this core only shows a tiny fraction of the Carboniferous shale which you might remember is thousands of metres thick in some places.

Just to the right in the diagram are other readings and measurements showing the amount of carbon in the shale (in percentage of carbon) and a sophisticated measure of the ratio of carbon isotopes in the organic matter of the shale ($\delta^{13}C$). The geologist has coloured the upper part of the core in light blue, and the lower part in browns and greys based on some very detailed sampling of the shale. The upper blue half, if you look carefully, has higher percentage levels of carbon (mostly over 2% and sometimes over 5%). So this would perhaps make the upper part of the shale better to drill.

Now I've studied this shale and it all looks the same to me from top to bottom. This might sound a bit negligent coming from a geologist, but in fact it's very difficult to distinguish different types of shale by eye. It's just too fine grained. You need a high-powered microscope to distinguish the minerals and organic matter in the shale, and specialist equipment to measure percentage carbon and $\delta^{13}C$. So you could never do this kind of analysis for every well that's drilled.

The key is to be able to understand why the upper half of the shale formed and why it's different from the shale in the lower part. To understand this you have to realise that the shale layer at the bottom is older and was deposited before the upper shale layer. At about the boundary between the blue and brown layers, the sea probably rushed in and a coastal region of shallow water close to wooded land suddenly became much deeper. This meant that there was more marine algae and plankton in the water and therefore in the mud. This makes the upper layer more attractive to drill. So if you can work out why the sea level rose and where it happened elsewhere in the north of England you might be able to plot the position of the upper shale type across a wide area and estimate its depth. This is the layer that you might drill for.

There's another reason why the upper shale type might be good to drill for. This is another kind of sweet spot.

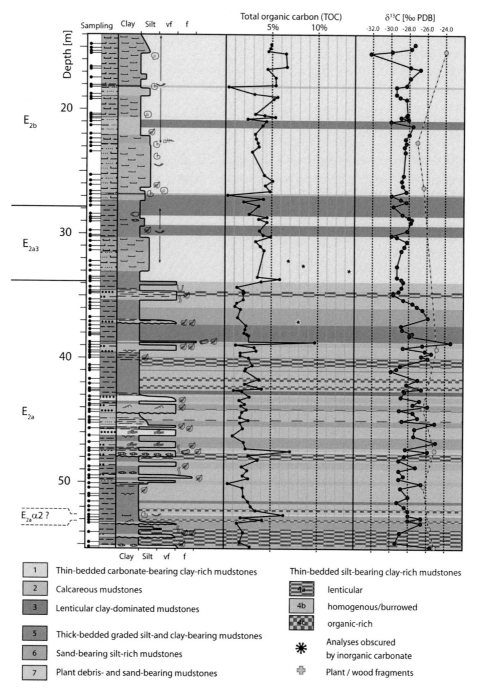

Figure 3.11 Layers of shale in a core from a borehole near Derby in Britain. On the left is a column with symbols showing the variation in the shale rock over 55 m. The other coloured columns show fossil content, and aspects of the chemistry of the shale. *Courtesy of Sven Könitzer et al. (2013) from his unpublished PhD thesis.*

Some shale layers respond better to fracking than others. You might remember that the high-pressure frack fluid is being squirted into rock to widen old cracks and make new ones. Surprisingly, shale can be quite flexible though, particularly if it contains certain minerals. Trying to frack rather 'plastic' shale might not lead to many cracks.

If we return to Fig. 3.11, you'll notice that the upper blue half is described in a couple of places as 'calcareous' or 'carbonate-bearing'. This means it's rich in calcium carbonate minerals like calcite. In fact some of this shale is cemented by calcite. The tiny particles are held together by calcite that has grown amongst them, perhaps long after the mud (shale) was deposited. This makes the upper shale type quite compact and brittle and probably easier to frack. In the last chapter, I mentioned that the Texas Eagle Ford shale had become a very attractive prospect and this is for the same reason. Some of the Eagle Ford shale has carbonate content as high as 70%. This and the low clay content make the Eagle Ford shale more brittle and easier to fracture. So it yields more gas and even oil. Perhaps the upper blue layer in Fig. 3.11 would frack better because it's brittle.

BIBLIOGRAPHY

King, G.E., 2012. Hydraulic fracturing 101: what every representative, environmentalist, regulator, reporter, investor, university researcher, neighbor and engineer should know about estimating frac risk and improving frac performance in unconventional gas and oil wells. In: SPE Paper 152596, Presented at the SPE Hydraulic Fracturing Technology Conference, the Woodlands, TX. 6–8 February 2012.

Könitzer, S.F., Davies, S.J., Stephenson, M.H., Leng, M.J., 2014. Depositional controls on mudstone lithofacies in a basinal setting: implications for the delivery of sedimentary organic matter. J. Sediment. Res. 84, 198–214.

Könitzer, S.F., 2013. Primary Biological Controls on UK Lower Namurian Shale Gas Prospectivity: Understanding A Major Potential UK Unconventional Gas Resource. Department of Geology, University of Leicester. Unpublished PhD Thesis.

Li, X.-Y., Liu, Y.-J., Liu, E.S., Feng, Q., Li, S.Q., 2003. Fracture detection using land 3D seismic data from the Yellow River Delta, China. The Leading Edge 680–683.

The Royal Society and The Royal Academy of Engineering, June 2012. Shale Gas Extraction in the UK: A Review of Hydraulic Fracturing DES2597.

Zoback, M., Kitasei, S., Copithorne, B., 2010. Addressing the Environmental Risks from Shale Gas Development. Briefing Paper 1. Natural Gas and Sustainable Energy Initiative. Worldwatch Institute.

CHAPTER 4

Gas in Our Water?

Contents

Lots of people the world over rely on water from rocks (groundwater), and they are naturally concerned about contamination by shale gas. Shale is a very impermeable rock that doesn't permit fluids to flow through it – that's why it has retained its shale gas rather than lost it. So it's difficult to imagine that gas could get through deep shale layers into shallow layers containing groundwater ('aquifers'). But could the fractures that are created by fracking fast-track the gas to the aquifers? Fractures like these aren't long enough to reach close to the surface but could they reach other natural cracks in the rocks, or old unsealed forgotten wells? This is unlikely if the engineers plan their wells properly. What's more likely is leakage from the shale gas well itself – from the cement and steel that line the well – so that gas bubbles up between the layers of cement and steel or between the lining ('casing') and the rock. There is clear scientific evidence that this kind of leakage has happened and that drinking water has been contaminated – though in a small percentage of wells. Wells can be fixed and contamination can be removed, but no one wants environmental damage of this kind whether it's reversible or not.

Keywords: Casing; Contamination; Fracking; Groundwater; Isotopes; Methane; Pennsylvania; Water cycle.

The term 'groundwater' is quite self-explanatory: water in the ground. We usually get it by digging wells, but sometimes groundwater comes out of the rock naturally in springs. Some parts of the world rely on groundwater for domestic use, agriculture and industry and dependence on groundwater varies a lot from country to country and within countries. For example in Sweden most water comes from the rivers, but in Denmark nearly all water comes from the ground. For obvious reasons people who rely on groundwater get very nervous about contamination. So one of the strongest concerns voiced by people over shale gas and fracking is: *will they contaminate my water supply?*

Shale gas and fracking
http://dx.doi.org/10.1016/B978-0-12-801606-0.00004-2

73

To answer this question we need to understand something about how groundwater gets into the ground and the form it takes below the surface.

Groundwater makes up about 20% of the world's freshwater. There's about the same amount of groundwater as there is freshwater locked up in permanent snow and ice at the north and south poles. This huge resource is very useful on its own and acts as an alternative source for surface water in times of drought.

Earlier in the book I explained how fluids – oil, gas, water – can be held in the spaces or pores between the particles of a rock. Sandstone – because it's made of quite large sand particles with big spaces in between – can hold a lot of fluid. It's permeable so it can let fluid in and it has space inside so it can store the fluid too, like a sponge. Geologists call it a reservoir, or a reservoir rock. But many other types of sedimentary rock can be reservoirs. When a reservoir rock contains water it's called an aquifer.

When rain falls on the ground it soaks into the soil and then, if the rock underneath is permeable, it can start a long journey downwards. The journey is part of the water cycle – similar to the rock cycle that I described earlier – but the water cycle describes how water moves around the earth (Fig. 4.1). Quite often the groundwater bit of the water cycle is ignored, much to geologists' annoyance, because while the water underground is out of sight, it is not out of mind – because it's so useful.

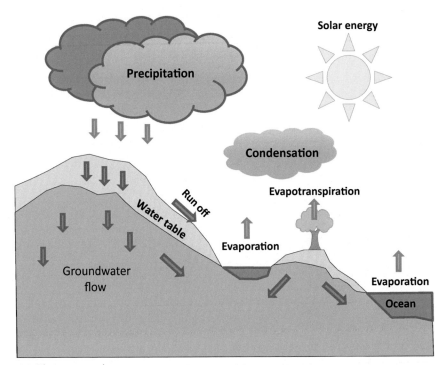

Figure 4.1 The water cycle.

Groundwater goes downward and settles there. It may move sideways but also it can stay put for a very long time, even millions of years. Water going downward eventually reaches rocks that are impermeable simply because the compression of the weight of rock above makes the spaces between the particles too small to allow water to get in. So water isn't found in rocks below a few thousand metres depth. Above this level, water can collect, soaked into the myriad spaces between the particles of sedimentary rock.

So the water collects and 'fills up' these rock layers. The mass of rock that contains water is called the 'saturated zone' and its top is known as the 'water table'. In places where the water table reaches the ground surface or intersects the ground surface, water pools or flows. It won't sink below because the rock underneath is already saturated with water (Fig. 4.2).

A simple analogy is when you dig a hole at the beach. The upper part of the hole and the surface of the sand that you're digging into are dry. Then you reach a depth in the hole which is suddenly wet. This level is equivalent to the water table. The sand underneath is saturated and water flows out and fills the hole no matter how hard you try to keep it dry.

Not all groundwater is useful. In fact most of it is too salty to use. Basically the deeper you go, the saltier the groundwater. The reason for this is mainly because the water has passed through a lot of rock on its way down and has dissolved salts. It's also likely to have been down there a long time. All aquifers under the seabed are also salty.

The most valuable aquifers are ones that are shallow (less a hundred metres) for ease of access and because they are more likely to be freshwater – but also because they are more likely to be refilled or recharged by rainwater. For this reason useful aquifers are likely to be far above any shale that might be fracked for gas.

FRACKING AND GROUNDWATER

There are three ways that fracking might interfere with groundwater. The first is a direct route whereby methane or fracking fluid travels through the fractures into an aquifer. This could happen by the fluid or gas simply following a very long man-made fracture or by fractures connecting with big natural faults or cracks in the earth – or even by a fracture connecting with an old oil or gas well that's been forgotten about. The second possibility is by fluid or gas leaking from the gas well straight into an aquifer on its long journey from the shale far below. A third possible route – usually discounted because of its very slow rate – is where methane molecules spread slowly through the groundwater like ink disperses in clean water – by diffusion. These are simply illustrated in Fig. 4.3.

Let's look at the first possibility. You might remember from earlier in the book that shale gas fracking can't take place in shallow rocks, mainly because there would not be enough pressure to force the gas out of the rock into the well. What this means is that fracking is always going to take place much deeper than freshwater aquifers. Probably there will be thousands of metres between the top of the fracked layer and the bottom of the freshwater aquifer. This distance is important. For contamination to occur either

Figure 4.2 The saturated zone and the water table. *Modified from http://water.usgs.gov/edu/earthgwaquifer.html.*

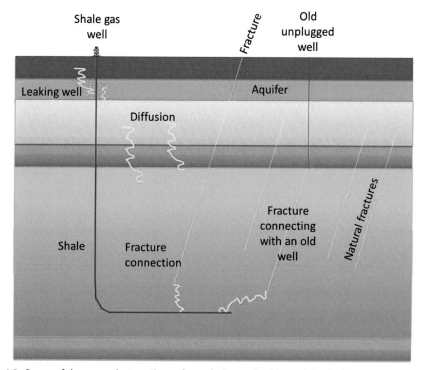

Figure 4.3 Some of the ways that methane from shale gas fracking might find its way into an aquifer. This could happen by the fluid or gas connecting with big natural faults or cracks in the earth – or even by a fracture connecting with an old oil or gas well that's been forgotten about. Fluid or gas can get straight from the well into an aquifer if the well is leaking. A very slow method of transmission could be diffusion.

man-made fractures or natural fractures would have to breach the gap between the shale and the aquifer. If *natural* fractures had breached that gap we would of course know because the aquifer would already have methane in it. In fact this isn't uncommon (as we'll see later in Pennsylvania) but such contamination is nothing to do with fracking.

Of far more concern is how far an *artificial* fracture can go underground. As luck would have it there's a scientific article about this very subject conveniently called *Hydraulic fractures: how far can they go?* It's by a team lead by Richard Davies, a geologist at Durham University in the UK, and was published in the scientific journal *Marine and Petroleum Geology*.

Before we plunge into the details of this article, I want to just digress into the world of scientific publishing because it is important for a lot of what you'll read in this book. There are thousands of scientific journals. Elsevier, the publisher of this book, is also the publisher of over a thousand scientific journals. So why are journal articles (or papers as they're known in the trade) so important? The first thing is that they focus on science topics very precisely without much extra stuff. The *Marine and Petroleum Geology* paper we're going to look at is mainly about man-made fractures. So it suits our interest. Second and much more important, journal papers are 'peer reviewed'. This means that the paper has been read and assessed before publication by other scientists in the field, usually anonymously. They've checked the data, the interpretation and the conclusions. Generally this is a very good check on the quality of the science – partly because scientists are a competitive lot and will generally be pretty critical of each other's work. In effect peer review is a gold standard. In a controversial area like fracking, peer review is vital to make sure that science isn't swayed or influenced and that it's reliable and credible. Of course it doesn't mean that bad science is never published, but more of that later.

We'll look at quite a few peer-reviewed papers in the chapters of this book, but let's get back to the article *Hydraulic fractures: how far can they go?* The authors compiled data from hundreds of separate fracking operations using microseismic monitoring as described in the last chapter. Amazingly these data can show the maximum limits of fractures. Most fractures open and grow vertically above and below the horizontal shale layer because it's easiest to open fractures in that direction (because the weight of rock from above makes it very difficult for horizontal fractures to open).

A graph of the American data is shown in Fig. 4.4. This is a complex chart but it shows a few things quite simply. Each coloured line represents data from different shale layers. The upper light blue line represents fractures in the Marcellus shale. The data are cumulative, showing many fracking operations at different depths. The 'fracture initiation depth' represents the level of the production casing where the frack was started from. The spikes in the same colour show how far vertically (the 'vertical extent' or VE) the different hydraulic fractures penetrated. The top horizontal line of the graph could be

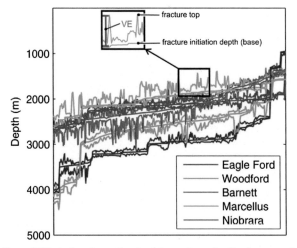

Figure 4.4 Compilation of US shale microseismic data on how far fracks spread from the horizontal well. *Compilation of data by Davies et al. (2012) from Fisher and Warpinski (2011).*

thought of as the ground level and so the graph also shows which are the deep and shallow shale layers in the US.

It's hard to read this off the graph, but it shows that the maximum VE of fractures in all the shale layers is 588 m in the Barnett shale, but most fractures are much shorter. The maximum VE in the Marcellus shale is 536 m.

Richard Davies and his colleagues used this study and a study of natural fractures to conclude that the probability of an artificial fracture extending vertically more than 350 m is about 1%. The maximum height of 588 m has been suggested subsequently as a minimum safe distance between the top of a fracking zone and the base of an aquifer, in other words no fracking should take place unless a shale layer is at least about 600 m below the aquifer.

It's worth thinking about this a little, based on what we've already learnt about subsurface pressure. As we found out before, once the frack has been completed the pressure in the well is considerably lowered and so the pressure difference between the inside of the well and the rock is reversed. This is why the gas and the flowback fluid rush out of the shale and into the well. At this point even if the fracture is full of methane and the fracture has grown unexpectedly far, the methane will most likely be travelling down the fracture towards the well rather up towards an aquifer.

A more troubling possibility is that a fracture opens and unknown to the drillers it grows and meets a nearby old well. You can be sure that if this did happen, it would be immediately apparent because of a sudden loss of pressure on the gauges at the surface. But by that time it would be too late, the methane and frack fluid would be on their way up to the surface through the old well. If the old well wasn't properly sealed off from the formations it passed through, it would contaminate the water in any aquifer on the way up.

To judge how likely this is let's look at another paper on this subject, again by Richard Davies and his team. This one's called *Oil and gas wells and their integrity: implications for shale and unconventional resource exploitation*. Part of this paper focuses on the possibilities that old unsealed wells might act as 'short circuits' to aquifers and to the ground surface. What they found was that there are a surprisingly large number of 'lost' or forgotten oil and gas wells. In fact around a million were drilled in America before there were formal regulations. In New York State alone where drilling has been going on for a very long time, around half of the wells drilled have no records. Some states (for example Texas) are spending a lot of money 'plugging' (or sealing) abandoned wells but inevitably wells that have been forgotten might never have been plugged. Just because a well is abandoned doesn't mean that it hasn't been plugged properly, in fact most probably are, but there is a risk.

I spoke to a senior well engineer recently about this very problem because exactly the scenario I've outlined above happened. The company engineers were fracking at high pressure. Suddenly the pressure dropped very rapidly and the engineers knew that the frack water was escaping somewhere. 'Somewhere' was a 5-m-high 'geyser' a few miles away in the woods. Deep below the surface a fracture had connected with an old unplugged well and the frack fluid was on its way up. As soon as this happened the pumps were shut down and the operation stopped. The company then spent a large sum of money plugging the forgotten well.

Despite spectacular events like this, the likelihood of this kind of short circuit is rather low. First the fractures have to be long enough to connect with an unplugged forgotten well. Most unplugged wells are likely to be older and much shallower than modern shale gas wells and so rogue fractures are unlikely to find them. Before fracking, the drillers will also look for nearby old wells. Quite apart from the environmental damage a leak like this would cause, it's also very expensive for the operating company.

The biggest risk of contamination by far is not forgotten, unplugged wells, but modern wells drilled expressly for the purpose of fracking. Put simply the main risk is that the layers of pipe and cement that seal the well from the rocks it passes through will leak somehow. You saw in the last chapter and in Fig. 3.6 how the drillers seal the wells: there are often two or three pairs of steel and cement layers at the shallow levels where the well is most likely to be passing through an aquifer. Can such a seemingly impregnable barrier be crossed? Can methane or frack fluid really get through the barrier?

Let's go through this in some detail. Remember that one of the chief reasons for casing a well is to stop it caving in. The steel tube of casing is pushed down the hole to support the sides. Then impermeable cement is pumped down the hole and forced between the outer surface of the steel tube and the rock. This seals the space between the rock and the steel tube. As we saw in the last chapter there can be several of these tubes and cement layers. If the well is operating, the main ways that leakage could happen is between the cement and the rock if the bond between them is loosened; between the cement and the steel tube; through the steel tube if it becomes corroded or through the cement if it becomes broken and degraded. If the well has finished its job and is plugged (usually with cement), the plug would have to leak as well.

Most countries require that engineering problems with wells be reported to the authorities. Do any of the above occur and how common are they?

The Department of Environmental Protection in the US state of Pennsylvania has a database showing problems with wells drilled into the Marcellus shale. The database can be interpreted in different ways and different investigators have suggested well problems in between 3.4% and 6.2% of wells over different periods for several thousand shale gas wells. The problems included very rare blowouts (uncontrolled explosions from the surface machinery of the well), and more common problems with casing. But these are engineering problems associated with wells, which may or may not have lead to contamination. What direct evidence do we have of contamination?

THE MYSTERY OF THE FLAMING TAP

Laboratory chemists are lucky because they can isolate things that might affect an experiment or an investigation, and eliminate them or compensate for them. Geologists working in the outdoors don't have that luxury. The earth is big and complicated.

What do you have to establish before you start looking for contamination of methane or frack fluid in aquifers? It's fairly simple. First you need to know what the precise condition of the aquifer was before any fracking took place – otherwise you can't tell whether fracking has caused any change. Going back to the flaming taps that you see in pictures and films about shale gas, you need to know if there's always been methane in water supplies or if it's a new thing. This isn't trivial. Over wide areas geologists don't know the natural amount of methane in aquifers or water supplies, simply because no one's really been interested enough to measure it before.

In Britain we've taken the decision to map how much methane there is in aquifers before shale gas fracking starts (if it starts). We call this a 'baseline' – the idea being that changes above a baseline might indicate that something is going wrong. The British Geological Survey is measuring amounts of methane and something of its character all across the parts of Britain that might in the future be developed for shale gas. As we'll see later, other countries haven't got a reliable baseline mainly because the shale gas business took off so quickly. This is a problem.

The second thing you need to know is that shale is not the only source of methane in rocks. I haven't mentioned this until now, but this seems a good point to introduce the fact that although shale makes methane underground, so do other processes. Living bugs in shallow soil and rock can make methane too. Just as methane pours out of buried rubbish tips or bogs and swamps so it can be made in shallow aquifers by bacteria and other organisms. I hesitate to introduce a pair of technical terms but they might be useful in the following discussion. The methane made in shale by deep heat is known as *thermogenic* for obvious reasons. The second type made by biological action in shallow rock layers and soil is known as *biogenic* methane. It's easy to use these words, but in practice it's quite difficult to distinguish between underground biogenic and thermogenic methane. This is important for scientific studies because biogenic methane is nothing to do with fracking and if you're studying methane in someone's tap water you have to know whether it's biogenic or thermogenic.

The best way to distinguish biogenic and thermogenic methane is a rather subtle technique using isotopes and other chemical ratios. Carbon occurs naturally in the form of three isotopes ^{14}C, ^{13}C and ^{12}C, and the ratios of the different isotopes in a carbon compound like methane are quite significant. As it happens, the isotopic ratio of $^{13}C:^{12}C$ (expressed as $\delta^{13}C$) in the carbon of methane is very different in biogenic and thermogenic methane. Without getting into detail, the $\delta^{13}C$ of biogenic methane is much lower than the $\delta^{13}C$ of thermogenic methane.

So there are two complications – natural amounts of methane and types of methane. With these in mind let's look at some published peer-reviewed studies of possible contamination during fracking.

The first is the most famous, published in 2011 by a team lead by Stephen Osborn of Duke University, North Carolina. The paper caused a huge stir – which is rare for a scientific article – and not only because it was published in one of the more august of the scientific journals, the *Proceedings of the National Academy of Sciences*. It was entitled 'Methane contamination of drinking water accompanying gas-well drilling and hydraulic fracturing' and contained in the summary the claim that there was *systematic evidence for methane contamination of drinking water associated with shale gas extraction*. Following the paper's publication, environmentalists against fracking were ecstatic because it provided the proof that they were looking for that fracking was damaging the environment. For the shale gas companies it was a disaster.

The clever thing about Osborne's paper was that the team chose a part of Susquehanna County in northern Pennsylvania where there were plenty of aquifers to sample (through water wells) and lots of recent fracking of the Marcellus shale. Some of the samples were taken around the town of Dimock where problems had been recorded before. Osborne's team measured the amounts of methane in the water wells and the $\delta^{13}C$ of the carbon in the methane. So they wanted to be able to say if there was a lot of methane (that could be ascribed to leaks from fracking, for example) and to say whether the methane was biogenic or thermogenic.

So far so good.

What did they find? Well you have a clue from the paper's title, but in essence they found higher methane concentrations in water wells close to shale gas wells than in wells further away. They also found that the $\delta^{13}C$ of the methane suggested that it was thermogenic and so couldn't be shallow gas generated by bugs. From this they concluded that the shale wells were leaking gas into aquifers.

Before we draw any conclusions, let's look at another paper in the journal *Groundwater*, this time by a team lead by the geochemist Lisa Molofsky. The paper looks at the same area that Osborn examined – Susquehanna County. It also re-examined some of Osborne's data. Though it's very technical there's a simple map in the paper that speaks volumes. It's reproduced here (Fig. 4.5).

The reason I like this map so much is that it shows one thing very clearly. There's a lot of methane in water wells in Susquehanna County! The red dots are sampled water wells and the size of the dot shows how much methane was found. There are red dots all over the map. It's also interesting – though not evident from the map – that there is methane in water wells whether they are near fracking operations or not. The second thing you can see if you look carefully is that the amount of methane in water wells seems to relate to the landscape – in other words there are high methane concentrations

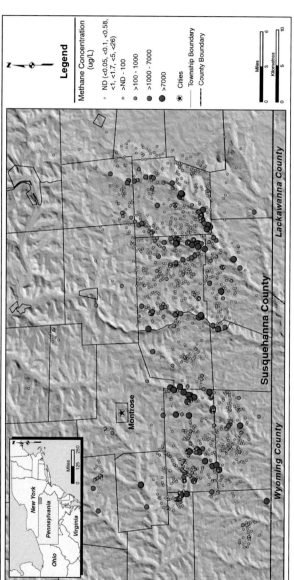

Figure 4.5 Elevation map showing the hills and valleys of Susquehanna County as well as dissolved methane concentrations from 1701 sampled water wells. *From Molofsky et al. (2013).*

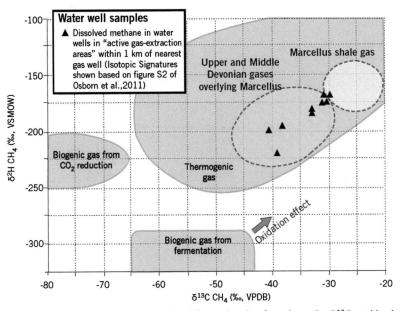

Figure 4.6 A scatter graph used to distinguish different kinds of methane by $\delta^{13}C$ and hydrogen isotope ratios in methane. *From Molofsky et al. (2011).*

in river valleys. Molofsky points this relationship out very clearly but also describes historical records that show that methane has been noted in Susquehanna County for over 200 years, long before fracking took place. There are even records of flammable and effervescing natural springs and water wells dating back to the late 1700s.

In another report by Molofsky published two years before in 2011, the data used by Osborn were analysed in a cross plot (Fig. 4.6). As it happens, the $\delta^{13}C$ of the carbon in methane can not only be used to distinguish thermogenic and biogenic methane, it can also be used to distinguish thermogenic methane from different shale layers. In fact the Marcellus shale in Pennsylvania is at the bottom of a great pile of shale layers over a kilometre thick. Each layer has gas with slightly different $\delta^{13}C$. These upper layers of shale are not quite as productive as the Marcellus shale so haven't been directly drilled for shale gas.

The methane samples in the scatter graph (black triangles) appear in an area indicating shale layers above the Marcellus shale. So this evidence suggests that there is thermogenic shale gas in water wells but that it's from layers above the Marcellus shale layer. But only the Marcellus shale is fracked in the area!

So the plot thickens.

You're probably thoroughly confused now but in fact a story is emerging. As Molofsky has shown so clearly, there is a lot of methane in Susquehanna County water wells. It even comes to the surface in springs naturally far away from fracking, and has done so for at least two centuries. It appears to be thermogenic (not biogenic) and so *it is shale gas*. It's coming up along river valleys more than elsewhere. Some of the samples indicate that the methane isn't coming from the Marcellus shale but from layers of shale above – which have not been fracked for shale gas.

What does it all mean? Probably it means that shale gas is coming up naturally along faults and cracks. These weaknesses in the rock are where rivers tend to flow because that's the easiest course for them to take. That might be why there are higher concentrations in water wells in valleys. The gas leaks into aquifers and also gets to the surface where it leaks into the atmosphere. Much of Pennsylvania water well gas is natural.

So are shale gas companies off the hook? Not at all. There's still the point that Osborne made originally – that methane is higher in wells near fracking areas. The 2011 paper has been criticised a lot – because its sample size is small and because there was no effort to understand methane in groundwater before the fracking. But Osborne's group from Duke University came back in 2013 – this time headed up by the geochemist Robert Jackson – and they altered their method to look at more samples and to examine the positions of wells in relation to river valleys. They came up with statistically reliable evidence that average methane concentrations were six times higher for water wells within 1 km of shale gas wells.

In the paper, Jackson's team suggested that the methane was probably leaking from poorly constructed wells where they passed through the aquifer (Fig. 4.7). They also mentioned the rather poor record of well construction in Pennsylvania: for example in 2010 there were 90 violations for faulty casing and cementing on 64 Marcellus shale gas wells.

It's interesting that while the studies from the teams at Duke University have found 'stray' methane they've not found evidence for stray chemicals used in fracking fluids which would surely be associated with the methane if they were injected into the rock to extract the methane. This is a bit of a problem but it may be because methane, being a gas, is more mobile than the chemicals in frack fluid.

Elsewhere in the US the shale gas business appears to be keeping its house in order. In a big study on another shale layer – this time the Fayetteville shale in north-central Arkansas – the Duke University group analysed the water from 127 drinking water wells. The Fayetteville shale has been intensively fracked with 4000 wells drilled since 2004, but very low concentrations of methane were found in water wells and the $\delta^{13}C$ of the carbon in methane suggested biogenic, not thermogenic gas. So it is unlikely to be from deep shale.

But overall the evidence seems to suggest that shale gas wells can and do leak and that the leaking gas finds its way into groundwater. It's not understood why a worrying number of wells are leaking in Pennsylvania while none seem to be leaking in Arkansas.

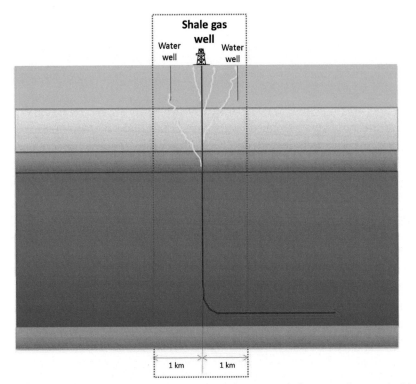

Figure 4.7 Robert Jackson's paper in 2013 suggested that some shale gas wells were leaking from their casings so that methane concentrations were six times higher for water wells within 1 km of shale gas wells.

SO IS IT SAFE OR NOT?

I'll examine the risks of shale gas overall in a later chapter, but groundwater contamination is an important enough issue to merit a discussion here. Embedded in the idea of how safe something is the concept of risk. We all take risks – like crossing the road or riding a motorcycle – and when we take risks, probably subconsciously we weigh up the risk. In weighing it up we're really balancing risk and reward. We're asking ourselves whether it's worth it. The same applies to shale gas. How likely is something to happen and what are the consequences of it happening?

We've seen some percentage figures for well problems. In Pennsylvania, for example, at one time or another 3.4–6.2% of shale gas wells have had problems. Some of these have obviously lead to water contamination. But in other areas thousands of shale gas wells have been drilled with no contamination.

What are the problems with methane in water? The main problem is that methane when concentrated in buildings or enclosed spaces is explosive, and there have been examples of explosions where natural or extracted methane has blown up. Methane becomes an explosive hazard at concentrations of 5–15% by volume in air.

The direct health effects of drinking water contaminated with dissolved methane are not well known, but the effects of chemicals used in fracking fluids could be very serious when concentrated.

Clearly shale gas and fracking are not without risk. What can be done to minimise the problem to begin with? In a paper in the journal *Environmental Science and Technology*, Avner Vengosh of the Duke University team recommends that new shale gas wells should be at least 1 km from the nearest water wells. They also suggest very careful collection of baseline data on the chemistry of groundwater and of gases from different layers in the rock that is being drilled through, and then continuous monitoring so that if there is a leak it will be possible to tell between gas from fracking and natural methane, and to immediately identify any chemicals from frack fluid. Then the operation can be shut down.

But if that has happened, can a leaking well be fixed and can a contaminated aquifer be cleaned up? The answer to those questions is yes – but no one wants the situation to get to that stage because environmental damage has been done, whether it is reversible or not. The companies do not want this either because the cost of fixing the problems will fall on them, as will the fines for bad practice. And no company wants the 'reputational damage' of being associated with an environmental incident.

But I haven't answered the question. Is it safe or not? Is there a risk? Well, the scientific evidence shows that there is a risk, but the risk can be minimised. It's more instructive to look at the risks of shale gas in relation to its benefits in the round, which is what I do in the last chapter.

BIBLIOGRAPHY

Davies, R.J., Almond, S., Ward, R.S., Jackson, R.B., Adams, C., Worrall, F., Herringshaw, L.G., Gluyas, J.G., Whitehead, M.A., 2014. Oil and gas wells and their integrity: Implications for shale and unconventional resource exploitation. Mar. Petrol. Geol. 56, 239–254.

Davies, R.J., Mathias, S.A., Moss, J., Hustoft, J., Newport, L., 2012. Hydraulic fractures: how far can they go? Mar. Petrol. Geol. 37, 1–6.

Fisher, K., Warpinski, N., 2011. Hydraulic Fracture-Height Growth: Real Data. SPE paper 145949.

Jackson, R.B., Vengosh, A., Darrah, T.H., Warner, N.R., Down, A., Poreda, R.J., Osborn, S.G., Zhao, K., Karr, J.D., 2013. Increased stray gas abundance in a subset of drinking water wells near Marcellus shale gas extraction. Proc. Natl. Acad. Sci. U. S. A. 110, 11250–11255.

Molofsky, L.J., Connor, J.A., Farhat, S.K., Wylie, A.S., Wagner, T., September 2011. Methane in Pennsylvania water wells unrelated to Marcellus shale fracturing. Oil Gas J. 54–67.

Molofsky, L.J., Connor, J.A., Wylie, A.S., Wagner, T., Farhat, S.K., 2013. Evaluation of methane sources in groundwater in northeastern Pennsylvania. Groundwater 51, 333–349.

Osborn, S.G., Vengosh, A., Warner, N.R., Jackson, R.B., 2011. Methane contamination of drinking water accompanying gas-well drilling and hydraulic fracturing. Proc. Natl. Acad. Sci. U. S. A. 108, 8172–8176.

Vengosh, A., Jackson, R.B., Warner, N., Darrah, T.H., Kondash, A., 2014. A critical review of the risks to water resources from unconventional shale gas development and hydraulic fracturing in the United States. Environ. Sci. Technol. 48, 8334–8348.

Warner, N.R., Kresse, T.M., Hays, P.D., Down, A., Karr, J.D., Jackson, R.B., Vengosh, A., 2013. Geochemical and isotopic variations in shallow groundwater in areas of the Fayetteville shale development, north-central Arkansas. Appl. Geochem. 35, 207–220.

CHAPTER 5

Did the Earth Move?

Contents

Fracking does cause earthquakes of a size that can be felt, but extremely rarely. Of the thousands of fracking operations that are carried out every year, only a handful have ever caused earthquakes large enough to have been felt. The key to the earthquakes are faults that are very common below the surface. These are natural fractures whose sides slip against each other when enough force is put on them. This slippage is very rare in most parts of the world where the faults are held tight by friction, but if tectonic and mountain building forces build up they may be too big to be resisted. The connection with fracking is that faults slip slightly more easily when there's frack fluid around so the engineers and geologists have to avoid faults – or if that's unavoidable – monitor activity so that they can shut the frack fluid pressure off if earthquakes are building up.

Keywords: Earthquakes; Fault; Fracking; Induced seismicity; Tremor; Water disposal.

On April 1, 2011, late on a Friday night, a small earthquake rumbled under the town of Blackpool in the northwest of England around 10 hours after an experimental frack of the nearby newly drilled Preese Hall well. Britain is an area of low earthquake activity – or seismicity as it's known – and Blackpool is a particularly quiet area. To find an earthquake of similar size in the area you have to go back to 1970. So the event though felt by very few people got noticed after it got into the news. Apart from the fact that earthquakes are rather rare in the area, it was also a big surprise for the company doing the drilling – who were not aware at the time that it had happened – because in fact earthquakes directly due to fracking are extremely rare.

A few months later (27th May), another smaller earthquake happened in the same place, again after around 10 hours after another frack in the same well. Following this second earthquake, the company stopped its operation, not to continue for another year. The two events, though too small to cause damage at the surface, had caused lots of local disquiet and national concern. It was largely through the *Blackpool earthquakes* that Britain first learnt about shale gas and fracking (Fig. 5.1).

Shale gas and fracking
http://dx.doi.org/10.1016/B978-0-12-801606-0.00005-4

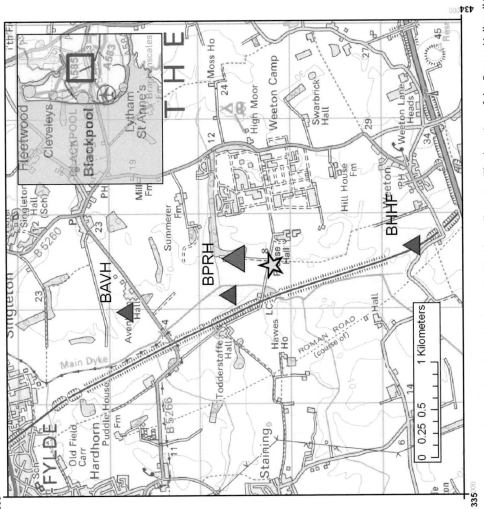

Figure 5.1 The epicentre of the two Blackpool earthquakes is marked by the yellow star. The location of the Preese Hall well is shown by the blue triangle. The red triangles show locations of temporary seismometer monitoring stations installed by the British Geological Survey. British Geological Survey data © NERC. Contains Ordnance Survey data © Crown Copyright and database rights 2014. Licence No 100021290. *From the BGS website. http://earthquakes.bgs.ac.uk/research/earthquake_hazard_shale_gas.html.*

How big were the two earthquakes? Seismologists used the following figures for them: 2.3 and 1.5 ML, both on the Richter scale. The ML unit is a measure of the local size or magnitude of the earthquake and it's really an estimate of the energy released by the earthquake which is the best measure of its destructive power. Of course whether you feel the earthquake also depends on how deep it is. Earthquakes can't be felt easily if they are under 3 on the ML scale and if they originate at typical fracking depths of say 2–3 km. So the two Blackpool earthquakes were tiny in comparison with many others, including many that occur in Britain because of the collapse of old coal mining tunnels (Fig. 5.2).

But there were no coal mining tunnels under Blackpool, so how and why did the earthquakes happen? How could fracking have caused them?

The first thing to realise is that the earthquakes weren't the direct result of fracking – in other words the rumbling detected by the seismometers at the surface wasn't from fracking. You might remember from an earlier chapter that the tiny shocks that result from the fracking itself are known as microseismic and these shocks are used by companies to follow the progress of the fractures. These are usually below 0 on the ML scale.

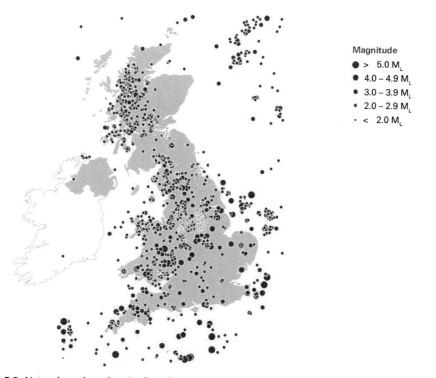

Magnitude
- ● > 5.0 M$_L$
- ● 4.0 – 4.9 M$_L$
- ● 3.0 – 3.9 M$_L$
- · 2.0 – 2.9 M$_L$
- · < 2.0 M$_L$

Figure 5.2 Natural earthquakes (red) and earthquakes related to coal mining activity (green) in the United Kingdom from 1382 to 2012. *From Royal Society and Royal Academy of Engineering Report (2012); data from the British Geological Survey.*

The fact that both the earthquakes happened 10 hours after the two main fracks (in other words after the maximum pressure of the pumps) gives us a clue to their origin: a fault — and probably the same fault each time.

FAULTS AND EARTHQUAKES

In this book we've only looked at faults in passing but they deserve some more attention. A fault is a crack or fracture in rock along which some movement has happened. Most earthquakes are caused by one side of a fault slipping against another.

Faults are very common and geologists on holiday with their families are often wont to bore their husbands, wives or children by pointing out faults in the landscape. Faults can often be seen in road cuttings or cliffs. One of the most spectacular is in the cliffs of Somerset in southern England (Fig. 5.3). Though the actual position of the fault isn't obvious its effects are: two distinct rock types are side by side.

In the next photo from a road cutting (Fig. 5.4), the exact position of the many faults is much more obvious, though the rocks affected aren't so different from each other.

In the picture there are two larger faults and several smaller ones. The larger ones have displaced the pinkish layer down to the right. Both the larger faults run diagonally up to the left. There are faults like these at the surface but of course they're underground too. You can see faults on the seismic section in Fig. 5.5 because of the displacement they cause, the most obvious one affecting the arch-shaped structure in the centre.

Figure 5.3 A large fault in the cliffs of Somerset, Southern England. The fault runs diagonally up to the right dividing the ochre coloured rock layers from the grey rock layers to the left. *From Wikipedia http://en.wikipedia.org/wiki/Fault_%28geology%29#mediaviewer/File:The_Blue_Anchor_Fault_-_geograph.org.uk_-_2455274.jpg.*

Figure 5.4 A road cutting exposing a cliff face with several faults displacing a pinkish rock layer down to the right. *Courtesy Sam Holloway and Gary Kirby. Photograph reproduced by permission of the British Geological Survey © NERC. All rights reserved. BGS copyright NERC.*

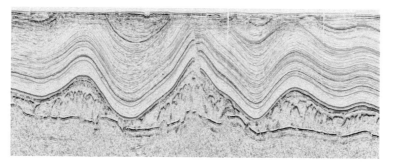

Figure 5.5 A seismic section of the rocks under the southern North Sea. A fault has affected the central arch-shaped structure. *Courtesy of WesternGeco.*

Although faults are very common, movement along them is very rare. For example the faults in the photographs probably haven't moved for millions of years, and might not move again for millions of years. They might *never* move again. If you're interested enough to get up close to a fault like the ones in the pictures, you'll often find the crack itself is filled with minerals and sealed up probably because the fault hasn't moved for a very long time.

Why do rock layers separated by faults move against each other and why has the pinkish layer dropped down to the right in Fig. 5.4? The main reason is that at one time or another the rocks of the earth's crust are put under compression or tension. Often it may be to do with big earth movements like plate tectonics or mountain building – or

their distant effects. These forces are too powerful even for rocks to resist and they fracture and shear in characteristic ways.

Sometimes faults want to move but can't. This sounds odd but it's because of the friction between the planes in contact on either side of the fracture. If you press your hands
together, palm to palm but at the same time slide one palm against the other you'll see what
I mean. Your hands want to slide but the friction between them stops the movement – until
the friction is overcome and then your hands slide quickly. This quick movement is in
essence an earthquake. The same happens deep in the earth – where a fault already exists
and where there's a force that wants to make the two sides of the fault move, the fault will
not move until the friction is overcome.

You've probably guessed that a fluid could make that movement easier; essentially this
happens because the fluid loosens the two sides of the fault as well as lubricating it. So
frack fluid could help to make faults move – and cause earthquakes.

Let's go back to the Blackpool example. A rather technical diagram is reproduced
below, from a report about the Blackpool earthquakes (Fig. 5.6). This deserves a bit of study.
The diagram shows the volume of frack fluid (mainly water) injected in the fracking and
the volume of flowback coming back after the fracking. It also shows the magnitude of the
earthquakes detected. It shows the 2.3 and 1.5 ML earthquakes, but it also shows that there
were a lot of much smaller earthquakes too. The yellow line showing water injected jumps
in five places and these are the five separate fracks that were carried out on the Preese Hall
well. The two biggest fracks (by water consumed) were accompanied by quite a few small
earthquakes. So there's clearly a relationship between water injected and earthquakes.

As I mentioned at the start of this chapter, the two main earthquakes happened
10 hours after each frack – though it's not obvious from the diagram. This may have been
the period of time it took for injected fluid to find its way to a small fault. Recently

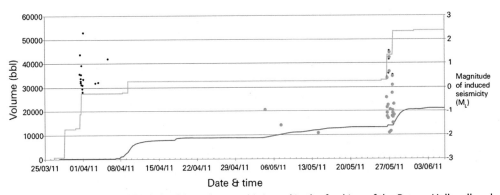

Figure 5.6 The volume of frack fluid (mainly water) injected in the fracking of the Preese Hall well and
the volume of flowback coming back after the fracking. The yellow line showing water injected jumps
in five places and these are the five separate fracks that were carried out on the Preese Hall well. The
red line shows the volume of flowback water. *From Royal Society and Royal Academy of Engineering
(2012), modified from de Pater and Baisch (2011).*

careful study of seismic sections has shown where this small fault might be – quite deep down but near the Preese Hall well. This fault was probably 'lubricated' by the fracking fluid so that it moved – at least twice. A magnitude 2.3 ML earthquake at this depth would equate to a movement between the sides of the fault of about a centimetre over an area of about a soccer pitch. To put this into perspective, the tiny microseismic vibrations used to monitor the progress of the frack are at their biggest about *minus* 1.6 ML and each one of these would be produced by a movement of a fraction of a millimetre – perhaps the width of a human hair – over an area of a square metre.

But not all faults will move if you pressurise them. If the force that created them has gone away then it doesn't matter how much they are pressurised, they won't move.

Clearly what this means for companies that want to frack is that they have to know where the faults are and whether there is some stress in the rocks that could make the faults move. Both of these are quite possible to do. Finding faults means studying seismic sections like the one in Fig. 5.5 very carefully. There are ways of working out levels and directions of local stress (any tension or compression) in rocks too.

But what if all the work is done, the fracking begins and earthquakes start – not the microseismic ones that are so useful, but the bigger ones? Well the answer is that the engineers will have to shut down the pressure. After the Blackpool earthquakes, the

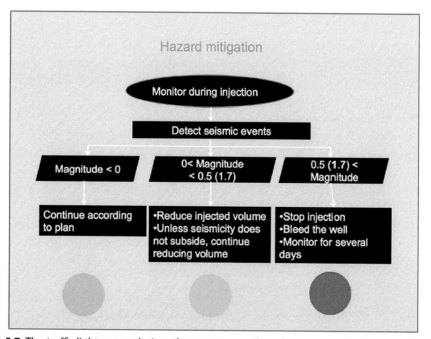

Figure 5.7 The *traffic light system* designed to manage earthquakes triggered by fracking. Essentially fracking can continue if earthquakes remain below 0 on the ML scale. If earthquakes rise in energy to between 0 and 0.5 ML then the amount of frack fluid injected must be reduced. If they go above 0.5 ML then injection must stop. *From Green et al. (2012).*

British Government following consultation with experts, came up with checks for fracking which have become known as the *traffic light system* (Fig. 5.7). If earthquakes rise in energy to between 0 and 0.5 ML then the amount of frack fluid injected must be reduced. If they go above 0.5 ML then injection must stop.

WATER DISPOSAL AND EARTHQUAKES

Earthquakes directly caused by fracking are extremely rare. Apart from the earthquakes I've described here, only a handful of any size that could be felt by people have ever happened across the world. Of far more concern are earthquakes caused by disposal of flowback and other water that bubbles up in oil and gas drilling of all kinds. This has produced more frequent and much larger earthquakes in the past.

The United States is the best place to study this. With lots of hydrocarbons (shale gas and conventional oil and gas) on land far from rivers and the sea, the industry has a problem with all the dirty water it produces. Flowback water from fracking you know about, but conventional oil and gas wells often gush dirty or saline water as well as the useful oil and gas. This is known as *produced water*. In arid areas or where there are no facilities to treat this water for surface disposal, the industry just puts it back down under the ground, in what are called in the United States *Class II disposal wells*. There are over 144,000 of these wells which together inject billions of litres of dirty water every day into deep rock layers.

There are over 200 water disposal wells in the Barnett shale area in Texas and these are the primary method of getting rid of flowback fluid that can't be recycled. The town of Cleburne, Texas, was recently affected: it had had no earthquakes in its 142 year history until 2008 and 2009 when small earthquakes measuring around 3 ML started to occur. The water is usually injected into deep sandstone layers above or below the Barnett shale. Some of the wells were injecting more than 20,000 cubic metres of waste water from fracking every month. Prompted by this spate of earthquakes, Texas geophysicist Cliff Frohlich used a set of temporary seismometers to measure any earthquakes accurately. He wanted to compare the positions of earthquakes, water disposal wells and known faults to see if there was a connection. He published the work in the journal *Proceedings of the National Academy of Sciences*. Near many of these wells, earthquakes were indeed quite common, but in a large proportion they weren't. Frohlich concluded that the wells that were near faults, and particularly faults with stress, were the most likely to trigger earthquakes.

The few studies like these suggest that permanent underground disposal of waste water is much more likely to produce earthquakes than fracking mainly because the water injected is permanent rather than temporary and because very large volumes can build up underground.

WHAT IS THE RISK?

Earthquakes connected with fracking (or water disposal) happen if fluid gets to a stressed fault that 'wants to move'. The lesson is that fracking or water disposal has to avoid faults. This means the geologists and engineers have to have a pretty good idea of what's down there. If they make a mistake and earthquakes begin to build up then there are ways to stop them getting any worse – by stopping injection or fracking.

What are the maximum sizes of earthquakes we might expect if the earthquakes happen before the pressure is dropped? The experts think that water disposal earthquakes rarely get bigger than magnitude 5 ML. The local effects of an earthquake like this are serious damage to poorly constructed buildings and slight damage to modern buildings. According to the British Royal Society and Royal Academy of Engineering, a 3 ML earthquake is the most you can expect from fracking fluid and this is felt by people but is very unlikely to cause any damage.

One last thing. If you read the last chapter you may be thinking that well casing – the all-important barrier between the gas and the aquifer – could be damaged by an earthquake because the casing is much closer to the action.

You'd be right. Tests carried out after the 2.3 ML Blackpool earthquake showed that the casing of the Preese Hall well had become slightly deformed, deep below the surface at a depth of more than one and half miles. The depth was so great and the position of the casing at that depth (where it passed through shale) meant that there was no risk of contamination, but it shows that earthquakes could increase risk of leakage.

BIBLIOGRAPHY

Frohlich, C., 2012. Two-year survey comparing earthquake activity and injection-well locations in the Barnett Shale, Texas. Proc. Natl. Acad. Sci. 109, 13934–13938.

Green, C.A., Styles, P., Baptie, B.J., 2012. Preese Hall Shale Gas Fracturing: Review and Recommendations for Induced Seismic Mitigation. Department of Energy and Climate Change, London. http://og.decc.gov.uk/assets/og/ep/onshore/5075-preese-hall-shale-gas-fracturing-review.pdf.

de Pater, H., Baisch, S., 2011. Geomechanical Study of Bowland Shale Seismicity, Synthesis Report. Available from: https://www.gov.uk/government/uploads/system/uploads/attachment_data/file/48330/5055-preese-hall-shale-gas-fracturing-review-and-recomm.pdf.

Royal Society and Royal Academy of Engineering, 2012. Shale Gas Extraction in the UK: A Review of Hydraulic Fracturing. Report DES2597.

CHAPTER 6

The Shale Gas Factory

Contents

Most people are more worried about what goes on at the surface than deep underground. If you've never seen a drilling rig or a frack truck, it's hard to imagine what it might be like to live up close to a fracking operation. Fracking is noisy and in most landscapes it will be quite obtrusive. To prepare the ground for the rig means heavy machinery. Things the rig uses – water, proppant, chemicals – usually have to be brought in, and stuff that the rig produces has to be taken away. So there are always a lot of trucks around. Some of the stuff that comes up from the ground like radioactive gas has to be carefully managed, and the water needs of fracking, which are considerable, have to be balanced against other water needs. When all is finished and the gas has been got out, the landscape may be crisscrossed with access roads and the cleared spaces where rigs once stood will still be visible though eventually the landscape will look much like it did. Inevitably though, shale gas fracking on a large scale counts as industrialisation of the landscape which to many people will be unacceptable.

Keywords: Abandonment; Drilling; Flowback; Fracking; Industrialisation; Landscape; Radioactivity; Traffic; Well pad.

I was at a conference on shale gas recently where all the talk was of the *shale gas factory*. Not only is shale gas unlike conventional natural gas geologically in that you have to frack to get it out, but its systematic extraction is different also. To get the gas out and make money, the companies drill across swathes of land dividing it into a grid, and advancing across the grid hoping to maintain their rates of production. A single well might *drain* a square mile of shale of its gas, but the next square mile will need to be drained too. The rigs move inexorably across the landscape. The more wells you drill, the more gas you get. So it's a bit like a factory – the more you put in at one end, the more *product* you get out.

To many people, more important than what goes on underground is what happens at the surface. We've seen that fracking is an intensive process needing lots of materials

Shale gas and fracking
http://dx.doi.org/10.1016/B978-0-12-801606-0.00006-6

– water, chemicals and energy. It also produces a lot – waste and useful gas. How are these arranged at the surface and what does the *shale gas factory* mean for local people? Is it just irritating and noisy – or is it downright dangerous? On my visit to a fracking operation at Fox Creek, Alberta what impressed me was the skull-battering noise, and the tangle of pipes and pressure hoses. But outside the football field-sized area where it all happened, the primordial birch forest seemed serenely quiet. Is fracking activity confined to the immediate area of the well? What are the effects on the countryside and the towns around?

To understand what shale gas and fracking looks like at the surface we'll follow it through from start to finish, from starting a well to plugging the well at the end – or as it's known in oil and gas parlance: from *spudding* to *abandonment*.

To spud a well is to make the early preparations to drill. This means bringing a drilling rig into position, connecting it to power and supplying the materials that are needed. Lots of these materials will be standing around waiting to be used and so they'll have to be stored. All of this is done in an area called the well pad (Fig. 6.1).

For argument's sake, we can start with a completely rural area with woods and a stream (Fig. 6.2). The area of the well pad has to be cleared and levelled. To protect the ground underneath from any chemical spills, an impermeable liner is laid and storage tanks and pipes are installed. Trucks have to visit the site very regularly, so roads or tracks have to be cleared. It may be that the local stream might be used to supply water and that waste fluid will have to be stored, in which case a lined waste pit has to be constructed.

Although in Fig 6.2 I've only shown one well, a well pad may have up to 10 well heads (the 'head' is the machinery at the top of the well). The wells are all likely to be horizontal deep down and are focussed on different parts of the shale below. If the exploration and test fracking is successful then many more well pad sites might be established with a spacing of as little as a mile apart.

Figure 6.1 A well pad. *From Shutterstock © Jim Parkin.*

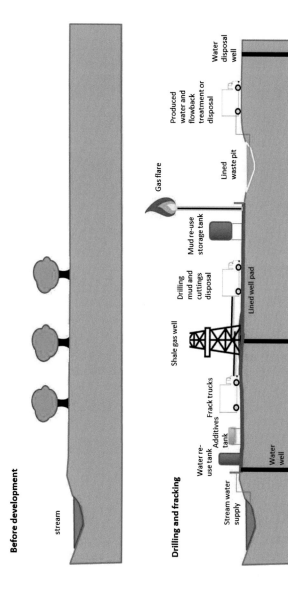

Figure 6.2 Before development, and during drilling and fracking – the most intense period in the well's life.

If fracking is successful, gas might be produced before the well head can be connected up to a national or local gas pipeline. This is difficult to store and so is often burnt in a flare which, though obviously wasteful of energy, is better than releasing methane, a much more potent greenhouse gas, directly into the atmosphere (more of this later). Other materials that have to be stored or disposed of, apart from waste fluid and gas, include drill cuttings (the bits of rock from the hole) and waste drilling mud (the lubricant used to keep the drill cool). These wastes are the same as those from conventional oil and gas wells. As I mentioned before, waste fluid is sometimes injected deep underground through disposal wells, especially in parts of the United States. Waste water and flowback fluid can also be transported to treatment plants especially set up to clean water from lots of industrial sources.

If the rate of flow of gas is high enough, the well will go into production, so it's connected to a national or local pipeline. This is the longest part of the life of the well that could stretch to many decades. The well might begin by producing up to 250,000 m^3 of gas per day in the first year declining to 10,000 m^3 per day after a decade, as the shale is drained. It's not unknown for companies to go back to 're-frack' the well, after perhaps 10 years, to increase production by opening up more fractures. Before gas can go into a national pipeline it has to conform to a standard and some machinery may still be needed at the well head, for example to remove water vapour from the gas. But production is much less visually obtrusive than drilling and fracking (Fig. 6.3).

When the well is no longer economic to operate, it's sealed up in a process called abandonment. This is another one of the terms that engineers use that is alarming to the public because it sounds a bit negligent! In fact abandonment is nowadays very strongly regulated. At the least, a cement plug is installed at the top of the shale layer, at the base of the lowest aquifer, and at the surface – and sometimes in other parts of the well too. These plugs stop any remaining gas getting up the well, into an aquifer or out to the surface. All in all, the final stage after the well is finished can look quite placid (Fig. 6.4).

WHAT DOES IT MEAN FOR ME?

Let's now examine what these stages mean for local people. What are the nuisances and inconveniences and what are the real dangers and risks?

You've probably guessed that the early stages are by far the most intense – and busier than in conventional oil and gas. More space is needed around the drilling rig because of the frack trucks, and access roads are more heavily used. The average size of a shale gas well pad is about 3 hectares whereas a conventional drilling rig makes do with 2 hectares. An interesting point, though, is that a shale well pad of a few hectares will be draining an area of shale below that's 10 times greater, because of the horizontal wells going out sideways. The well pad might be in use for up to 40 years and so the land is out of commission for any other use. After it's all finished though, the land can be used pretty much as it was before.

Production and processing

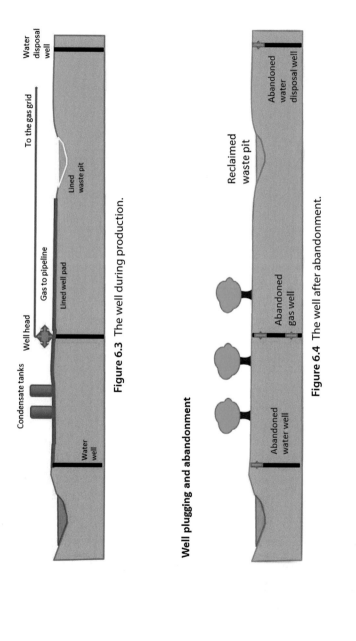

Figure 6.3 The well during production.

Well plugging and abandonment

Figure 6.4 The well after abandonment.

The noise is intense during the fracking and drilling, and will be a nuisance during excavation to prepare the well pad. Flares of burning gas look quiet but can actually be quite noisy. But many of these activities, particularly drilling, will be continuous for 24 hours a day. Drilling can continue for a month without stopping. It can be shielded to try to reduce noise but it's still far louder (if you are standing nearby) than most industrial activities that people would get close to, like building sites or factories. Noise during production is close to nothing, and after abandonment is zero.

If a well pad is established in wooded land most of the activity is hidden by trees – all except for the drilling rig which could be in position for a month or more, and for a multi-well pad many more months. Rigs can be 30 or 40 m tall. During production there's nothing much to see unless there's a flare. In open land, shale gas operations look pretty stark and do alter the landscape considerably, leading to local peoples' worries over the 'industrialisation' of the landscape. In the shale gas factory the countryside is divided up and drilled systematically so that access roads split the land into squares or rectangles (Fig. 6.5). The noise and the traffic at the most intense periods of drilling and fracking could have an effect on local wildlife and compartmentalise their habitat. The pipelines linking up the producing well heads could do the same. The trucks might bring in invasive species too.

And now to the trucks! These really are enormous. Each one carries enough engine to power a small village and many frack operations need trucks to bring in water and proppant (if you can't get it locally) – and to take away old drilling mud, drill cuttings and flow back water. You need trucks to excavate and level the well pad also. This means

Figure 6.5 Land divided up following conventional oil and gas extraction in northern Alberta, but land following systematic shale gas extraction looks similar. *Photo M. Stephenson.*

an enormous amount of truck traffic: between 7000 and 10000 single truck journeys have been estimated per well pad through the period of construction and fracking, though this would diminish very quickly when the well begins to produce commercially.

This number of trucks would hardly go unnoticed – there would be more trucks on public highways affecting traffic flow and increasing congestion. Large trucks on narrow roads are a hazard, and they damage roads and bridges (Fig. 6.6). Trucks carrying hazardous fluids sometimes crash or leak. Of course we shouldn't forget the emissions of the trucks – and the diesel that's burnt in the engines. I'll deal with the overall carbon emissions associated with shale gas and fracking later in the book, so I haven't forgotten!

But all of these are nuisances that are associated with many other industrial and oil and gas activities. I don't mean to minimise them, but they are the sorts of things that we can cope with. Trucks can be re-routed; noise can be put up with; land can be reclaimed just like it can after any industrial activity, like quarrying.

What are the risks peculiar to shale gas at the surface? There are two different kinds. One isn't really a risk, more a tension. This is the question over how much water is used in fracking. In conventional oil and gas, water isn't used in huge quantities; in fact a lot of hydrocarbon extraction actually produces more water than oil and gas. But you've seen that fracking uses enormous amounts of water, up to 250 litres of water a second for a single well – that's about a bath tub every second. This means that tens of millions of litres might be needed for a full frack. A hundred wells a year means two million cubic metres

Figure 6.6 A truck carrying proppant to a fracking site. Photo *M. Stephenson*.

(*not litres*) of water a year. Can we really spare this sort of amount of water? Well, it depends on where you are.

Compared to, for example, the licensed annual water extraction in England and Wales of 12600 million cubic metres, two million a year isn't much. But it needs to be seen in the context of how much water is needed and available in different areas of the country. In many parts of south eastern England for example, water resources are already stressed due to overextraction from reservoirs and aquifers. The same is true in parts of Texas, Oklahoma and Alberta. What happens if water is taken too much or too fast either from water wells or rivers? Rapid extraction during periods of low flow could affect fish and other aquatic life due to the waters' increased temperature and low oxygen concentration. Public water supply and water for agriculture could also be affected. With climate becoming more variable, and with drier summers, even more stress will be put on water resources.

In most parts of the world water abstraction is regulated so we would hope that where fracking would need more water than could be supplied, the drillers would not get a licence to take water. If they really need water, drillers could work towards recycling more of the flowback, or perhaps start looking at more plentiful waters source, like the sea.

The second problem I mentioned is quite different. You might remember from the first chapter that a New York newspaper reported that radioactive rock cuttings from drilling in the Marcellus shale were in danger of being disposed off in county landfill sites not designed for them. Can shale be radioactive?

Well, yes it can. Like many other rocks, shale is naturally radioactive, but at a very low level – less than for example granite. The reason why is rather interesting and to understand it we have to go back to the formation of shale, or the mud that it once was. This mud collected in deep water, far from land and being far from the source, the mud collected rather slowly, perhaps only a fraction of a millimetre of sediment thickness per year. It was mostly stuff falling slowly out of suspension, like fine coffee grounds collecting at the bottom of a cup.

Naturally radioactive dust and particles fall in a constant very low-level 'rain' in the atmosphere and the sea. Where the sedimentation rate is very slow, this 'rain' gets concentrated in layers that become more radioactive than others. These more radioactive layers just reflect the slow sedimentation rate. The clay minerals and organic matter in the shale is also good at binding to the radioactive particles and so the shale retains its radioactivity. But it's still very low – much lower than anything that could cause harm.

The diagram shows these radioactive elements (Fig. 6.7). The vertical coloured squiggles are records picked up from a tool that's lowered down the boreholes and wells to measure natural radioactivity in the rock layers. This one shows a single well for which

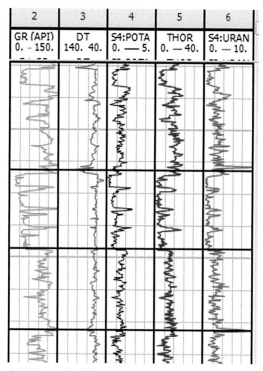

2	3	4	5	6
GR (API) 0. - 150.	DT 140. 40.	S4:POTA 0. — 5.	THOR 0. — 40.	S4:URAN 0. — 10.

Figure 6.7 Part of a borehole through shale showing concentrations of radioactive elements, including thorium and uranium. *Redrawn using data from Leeder et al. (1990).*

several radioactive elements have been measured, including uranium (the red squiggle). The measurements along the top are in parts per million. You can see that there are two thin levels within this borehole where shale has uranium levels of ten parts per million. But most of the shale contains much less. The granite of the southwest of England measures 10–20 ppm of uranium throughout. Shale also contains thorium and a radioactive isotope of lead called lead-210.

But you want to know whether this radioactive material can get out to the surface.

The most obvious way would be by flowback fluid. This stuff has been injected at very high pressure and has really 'got amongst' the shale deep down and then has come back to the well and been pumped back to the surface. Could it have picked up uranium, thorium or lead-210 on its journey?

These three are quite immobile, but uranium and thorium both decay to radium which is fairly soluble in water. Predictably studies have detected radium most often in flowback, and radium values from the Bowland shale of northern England and the Barnett shale in Texas are much higher than natural groundwater in the area, but still well below dangerous levels.

Radium also decays to radon, a radioactive carcinogenic gas, and radon is soluble in water and so will be present in flowback, dissolved and as a gas. Radon, being liberated at the well head for whatever reason, would likely disperse very quickly in the open atmosphere, but it could pose a problem in confined spaces like pipes. The main problem is the decay of radon which happens rather quickly (its half-life is 3.8 days). This decay produces radioactive isotopes such as lead-210, bismuth-210, polonium-210 and lead-206 in *solid* form that collect in gas pipes. It's known as 'scale' and it's a well-known problem in the oil industry because most natural gas contains radon, even conventional natural gas. So many gas pipes have to have the scale removed. This scale is radioactive enough to be disposed of in specialised sites. Regulations don't allow gas with high radon or pipes that are affected by the scale.

Is radon mobile enough to find its way directly into groundwater from shale deep below? Movement through diffusion is very slow, probably taking thousands of years by which time the radon will have decayed to solid radioactive isotopes which stay put. Even if it was forced through fractures its short half-life would probably prevent it moving far.

NUISANCE OR REAL DANGER?

I think it's clear that shale gas drilling, particularly in its early stages, will be unpleasant and certainly a nuisance if you live close up to it. There will be more trucks on the road, there will be more noise and the drilling rig operating 24 hours a day will be a blot on the landscape. If you live in a region where shale gas is big business, with hundreds or thousands of wells being drilled per year you will really notice a big difference because drilling will be going on somewhere most of the time.

Will these activities be hazardous? They might be. Trucks might spill chemicals; waste tanks might overflow in a storm. But these are industrial installations that engineers are good at managing and have been managing for a long time. In many ways they're no different from building sites.

As for industrialisation of the landscape and the shale gas factory, there's no doubt that for a period that could last for as much as a year there will be intense industrial activity. After this, during production, activity is less intense and obtrusive, and after abandonment there's no activity. Whether you think that the landscape is scarred and tainted with industry at this stage depends on your point of view. It's true that access roads will still divide up the land after the wells are plugged, and clearings in the woods will still be visible for a long time after. Some will say that that's what our landscape looks like already – a pattern of past uses of the land. Others will say it's unacceptable.

One thing we haven't covered is the wider environmental impact of shale gas beyond the local: the effect that burning shale gas for power might have on our atmosphere. This is the subject of the next chapter.

BIBLIOGRAPHY

AEA Technology, 2012. Climate Impact of Potential Shale Gas Production in the EU. Report for European Commission DG CLIMA, AEA/R/ED57412 67.

AEA Technology, 2012. Support to the Identification of Potential Risks for the Environment and Human Health Arising from Hydrocarbons Operations Involving Hydraulic Fracturing in Europe. Report for European Commission DG Environment AEA/R/ED 57281.

Almond, S., Clancy, S., Davies, R., Worrall, F., 2014. The Flux of Radionuclides in Flowback Fluid from Shale Gas Exploitation. Environmental Science and Pollution Research, published online 18 June 2014.

Leeder, M.R., Raiswell, R., Al-Biatty, H., McMahon, A., Hardmann, M., 1990. Carboniferous stratigraphy, sedimentation and correlation of well 48/3-3 in the southern North Sea Basin: integrated use of palynology, natural gamma/sonic logs and carbon/sulphur geochemistry. Journal of the Geological Society of London, 147, 287–300.

Public Health England, 2013. Review of the Potential Public Health Impacts of Exposures to Chemical and Radioactive Pollutants as a Result of Shale Gas Extraction. http://www.hpa.org.uk/webc/HPAwebFile/HPAweb_C/1317140158707.

Stuart, M., 2013. Water Supplies May Struggle to Cope with Fracking Demands. The Conversation. http://theconversation.com/water-supplies-may-struggle-to-cope-with-fracking-demands-17296.

CHAPTER 7

Shale Gas and Climate

Contents

This chapter looks at whether shale gas has a place in modern energy. Big energy – the energy that powers much of the modern world – comes from fossil fuel. It makes our cars and planes go and it generates most of our electricity. But burning it produces CO_2 that warms the planet. Advocates of shale gas think that burning it in power stations is much better than burning coal. They know it's not as low in emissions as using renewables to generate electricity – or for that matter nuclear – but they think it will be difficult to get to low emissions without some help, because we're so energy hungry. Probably we'll get even more electricity-hungry in the future because we'll want electric cars.

So shale gas looks like a convenient fuel to bridge the gap to a sunny upland of very low carbon energy. But is shale gas really a low carbon fuel? It might burn cleaner than coal, but what about the energy needed to get the shale gas out of the ground – all the diesel burned in the frack pumps – doesn't that negate the advantages of the 'clean burn'? What about the sinister 'fugitive emissions' of methane? Do they make gas worse than coal as some people think? Should shale gas play a part in world energy or should it be left in the ground?

Keywords: Carbon dioxide; Coal; Emissions; Fossil fuel; Fracking; Fugitive; Policy; Renewables.

The central tension with energy is that we want more and more of it, but we don't want the carbon dioxide that comes with most of it. By *most of it* I mean fossil fuels – the fuels that came from once-living things that have sat in the ground for millions of years. Shale gas of course is one of them.

A fairly constant message in the media over the last few years has been that shale gas is *better than coal* – better for the climate. The switch made in the US from burning a lot of shale gas to make electricity instead of coal has been seen as the reason for a reduction in US CO_2 emissions over the last few years.

But the idea of switching from coal to gas (whether shale gas or conventional gas), has a powerful scientific basis also. Two scientists from Princeton University, Stephen Pacala and Robert Socolow, published an article in the journal *Science* in 2004 with the rather

Shale gas and fracking
http://dx.doi.org/10.1016/B978-0-12-801606-0.00007-8

impenetrable title 'Stabilization Wedges: Solving the Climate Problem for the Next 50 Years with Current Technologies'. It was an interesting article though. Pacala and Socolow argued that we can solve the climate problem with the technology we've already got, and by being a bit less wasteful. They settled on the idea of stabilising carbon emissions at 7 billion tonnes per year. This, they said, would avoid the worst effects of climate change.

They pictured CO_2 emissions in a way that is quite useful for non-scientists (Fig. 7.1). The diagram shows carbon emissions from fossil fuels increasing until the present day and then two imaginary possibilities for the future. The bad one called the 'current path', which involves not doing very much about carbon emissions, would eventually produce a tripling of CO_2 in the atmosphere. This is also known as the *business as usual* scenario amongst some climate scientists. The other projection – the 'flat path' – is what we could achieve by lowering emissions and keeping them low.

So the yellow triangle between the current path and the flat path in the diagram is the carbon emissions we've got to avoid, the carbon 'savings' we have to make. This stabilisation triangle – as they called it – represents a huge amount of carbon, and so they split it up into a number of smaller *wedges* that might make the task more manageable (Fig. 7.2).

Each wedge is an activity that on its own – if done between now and 2055 – could stop a billion tonnes of extra carbon from getting into the atmosphere (Fig. 7.3). The nuclear fission wedge is simple: triple the world's nuclear electricity capacity by 2055 and we'll achieve a wedge. In wind, the installation of 1 million 2 megawatt windmills to replace an equivalent amount of coal electricity generation would produce a wedge. In solar, the

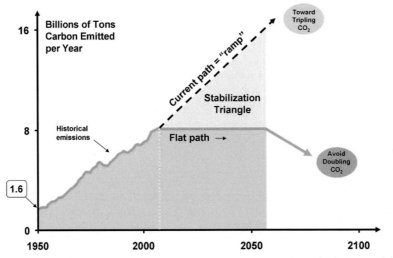

Figure 7.1 Carbon emissions from fossil fuel burning. Time in years is along the bottom of the graph. *From the Carbon Mitigation Initiative Website, Princeton University http://cmi.princeton.edu/wedges/. Courtesy of the Carbon Mitigation Initiative, Princeton University.*

installation of 20,000 square kilometres of solar panels by 2055 would produce a wedge. Perhaps simplest of all would be to eliminate tropical deforestation or plant new forests over an area the size of the continental United States. Both would produce a wedge.

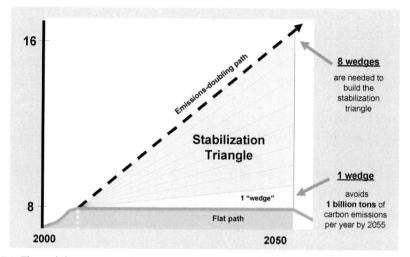

Figure 7.2 The stabilisation triangle. *Courtesy of the Carbon Mitigation Initiative, Princeton University.*

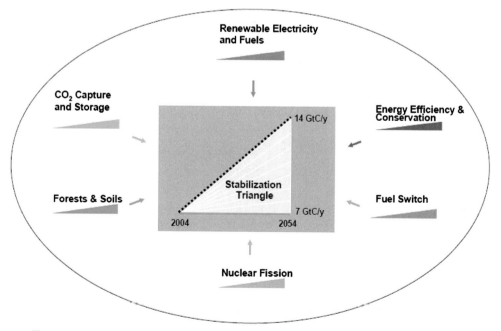

Figure 7.3 Types of wedges. *Courtesy of the Carbon Mitigation Initiative, Princeton University.*

For us the 'Fuel Switch' is the important one. The idea is to fix it so that more electricity is generated by gas power stations than their coal relatives. In their paper, Pacala and Socolow say that if 1400 natural gas power stations were substituted for an equal number of coal-fired power stations then this would save one wedge of CO_2 emissions. The assumption that underlies this is that gas makes as much electricity as coal but with less CO_2.

Let's look at this in more detail. To compare the emissions of different fuels, you have to look at the amount of CO_2 emitted per unit of energy or heat given out, not the volume of the fuel used. Obviously by volume, coal would burn to make much more heat than gas because coal is heavy, dense stuff. A useful unit to compare the amount of heat a fuel produces is the British Thermal Unit (BTU – often used in the form million BTU). If you take several fuels and compare the amount of CO_2 they make when burnt to produce one million BTU (the same amount of heat), you get the information in Table 7.1.

Put simply, CO_2 emissions per unit of heat are smaller in natural gas than coal, so unit for unit, electricity produced from a gas power station makes about half as much CO_2 as from a coal plant.

So is shale gas a low carbon fuel? In fact it's not as simple as CO_2 emissions from burning shale gas in the power station. You also have to think of the CO_2 emissions that might be associated with shale gas exploration and production – like the diesel-powered frack pumps. How much CO_2 gets emitted as part of the whole process and how does that compare with, say, coal mining? This survey of the emissions of the whole of the process is known as a lifecycle assessment.

Methane comes into it as well, though nobody knows for certain how much. Methane from a leaking well goes straight into the atmosphere. Many shale gas operations also deliberately release methane during drilling, or even release and burn gas in great flares during production when there's no way of collecting it for sale.

Table 7.1 The CO_2 emissions of different fuels

Fuel	Pounds of CO_2 emitted per million BTU of energy
Coal (anthracite)	228.6
Coal (bituminous)	205.7
Coal (lignite)	215.4
Coal (subbituminous)	214.3
Diesel fuel and heating oil	161.3
Gasoline	157.2
Propane	139
Natural gas	117

The unit is pounds (weight) of CO_2 per million BTU of energy.
From the US EIA, http://www.eia.gov/tools/faqs/faq.cfm?id=73&t=11.

Methane is much worse for the climate than CO_2. It doesn't last as long in the atmosphere as CO_2 (it slowly decomposes) so methane's effect gets lower with time. But while it's around, methane is much more efficient at trapping heat than CO_2 so it's worse for global warming. Pound for pound or kilogram for kilogram over a 100-year period methane is over 20 times worse than CO_2 for global warming.

This is where things get controversial again so we're going to look at a few peer-reviewed papers. Like many great scientific controversies, this one starts with a paper that very much challenges the normal way of thinking. It was by Robert Howarth and colleagues at Cornell University and it was published in the journal *Climatic Change*. Their startling conclusion was that over a 20-year period the effect of shale gas (its *greenhouse gas footprint* in the jargon) was 20% *worse* than coal, and maybe even *twice as bad* as coal. This seemed to turn the orthodoxy on its head and so the paper was heavily scrutinised. The gas companies didn't like it because it seemed to take away one of the plus points that shale might be cleaner than coal. The advocates of shale gas as a bridging fuel didn't like the paper. The coal companies were quietly pleased!

With what we know about the energy value and emissions of the two fuels, how could shale gas be worse than coal? The answer lies in the methane. Even though it may not leak methane very much, said the Howarth group, the small amount is very bad for the atmosphere because of the global warming power of methane. They thought that 3–8% of the total methane production escapes to the atmosphere through the lifetime of *every shale gas well* – from drilling, through fracking, production and abandonment. In other words 3–8% of the total methane doesn't actually get into a pipeline. This is enough leaking gas to really make a difference. Howarth and his team also said that fracking made shale gas wells leak much worse than their conventional cousins.

To understand their findings we have to go back to drilling. There are two parts of the shale gas process that Howarth thought were particularly guilty: the flowback and the cryptically named process of 'drill-out'. The first you know about – this is when the fluid that's been injected into the shale starts pouring back into the well. It might flow for days before a lot of gas comes out. But what I didn't tell you before is that flowback contains dissolved methane or bubbles of methane. The flowback comes to the surface and often just sits around in open tanks at the surface. The Howarth group thought that this let an enormous amount of methane into the atmosphere. They didn't have many direct measurements of actual methane in flowback (about 10 tests of wells drilled into the Haynesville shale), but those tests suggested that an average of 6800 cubic metres of methane were released per well during flowback. If we were talking about CO_2 as a global warmer this would be equivalent to more than 100,000 tonnes of CO_2! Howarth compared this amount to the 1000 cubic metres of gas that might leak from a conventional gas well *throughout its lifetime*.

What's drill-out and what's bad about it? You might remember the plug and perf method of fracking where bits of the well are separated off to subject parts of the shale to fracking pressure. The separation is done with plugs of cement or other strong materials. After all the sections have been fracked, the plugs have to be drilled through to release the gas to the surface. Often this gas can't be collected for piping to the grid at first and may be just released (or vented) to the atmosphere.

In what's called a *rebuttal*, another group from Cornell University published a criticism of Howarth's paper in the same journal, *Climatic Change*. The editors of journals rather like this kind of exchange and are usually willing to publish criticism of previous papers, as long as the criticism isn't personal and sticks to the subject. It means that the science they publish is active and that it matters. The rebuttal, by a team lead by Lawrence Cathles, mainly objected to the high leakage rates that Howarth used and suggested that the shale greenhouse gas footprint is half or a third of coal's, much lower than suggested by Howarth.

So part of this debate is about different measured amounts of escaping gas during drilling processes. Later studies and reports, for example, a report by David MacKay and Timothy Stone for the British Government listed the measured amounts of methane released by flowback fluid from several studies. This is summarised in Table 7.2. MacKay and Stone pointed out the large size of Howarth's figures – taken from the 10 wells in the Haynesville shale – in comparison with other figures.

Another way of measuring methane emissions from shale gas wells is to collect samples higher up in the atmosphere above regions where shale gas is produced. There have been several studies of this kind recently. One lead by Scott Miller in the *Proceedings of the National Academy of Sciences* used measurements from planes and towers, and a computer model of the atmosphere over the US, to estimate fugitive emissions. They used the geographical patterns and proportions of other gases mixed with the methane to suggest that methane emissions from north east Texas for example (around the area of the Barnett shale) are much larger than US government estimates made by the US Environmental

Table 7.2 Comparison of amounts of methane released from flowback from various shale gas operations in the US

Source (mainly scientific papers and reports)	Shale layer	Volume of gas released during flowback (thousands of cubic metres per well)
Jiang	Marcellus	603
Howarth	Haynesville	6800
Howarth	Barnett	370
US Environmental Protection Agency	Various	260
O'Sullivan and Paltsev	Haynesville	1180
O'Sullivan and Paltsev	Barnett	273
O'Sullivan and Paltsev	Fayetteville	296
O'Sullivan and Paltsev	Marcellus	405
O'Sullivan and Paltsev	Woodford	487

The sources listed are mainly scientific papers and reports.
Adapted from McKay and Stone (2013).

Protection Agency. Are these broad brush atmospheric measurements more reliable than the patchy measurements from actual well operations? Perhaps, but can we be sure that the aircraft measurements are attributing methane to the right sources, after all swamps and municipal waste dumps produce methane – as do cattle. And cattle are common in Texas!

So the controversy continues because plane measurements and modelling can't be unequivocally related to shale gas drilling, being too remote from the action. Direct measurements at the drilling sites run the risk of being unrepresentative if there are two few of them. Perhaps the answer is more numerous direct measurements so that emissions can be clearly attributed. A very recent paper, again in the *Proceedings of the National Academy of Sciences*, is on the right track. It's by David Allen and colleagues at the University of Texas at Austin and it describes direct measurement of 190 shale gas sites all over the US in an attempt to be representative of the main shale layers. Allen scaled up the results to suggest that the leakage rate is about half of 1% of gas production, much less than the 3–8% estimated by Howarth.

SO SHOULD SHALE GAS BE LEFT IN THE GROUND?

Nobody seriously wants to continue burning fossil fuels forever. Apart from their climate-changing potential through methane and CO_2, they are dirty in other ways – they emit sooty carbon to the atmosphere for example. Also coal, oil and gas are dangerous to mine or extract. Coal is particularly dirty and dangerous. When it's burnt it releases some unpleasant things that it contains in small amounts, so that they disperse in the atmosphere – like sulphur, cadmium and mercury. Hundreds of thousands of people have died in coal mining accidents: in the US more than 100,000 coal miners were killed in accidents in the twentieth century; in China in 2004 alone there were over 6000 deaths related to coal mining either in accidents or due to poor health.

So we don't want to continue using fossil fuels if we can help it. We don't want to use shale gas for too long either. The question is – can shale gas (or natural gas generally) be a bridging fuel? To see if this is possible we have to compare shale gas against the energy sources it might replace and the sources it might build a bridge to. I'll have a go at looking at shale gas and coal first.

Shale Gas Replacing Coal?

Most coal that's mined in the world is used to produce electricity in huge power stations. The US is no different: coal is by far the biggest source of American electricity but since shale gas arrived less coal has been burnt. In 2000, coal powered 52% of American electricity, and natural gas 16%. By 2013, the roles had almost reversed because of cheap gas (to 39% coal and 27% gas). Most of the cheap gas was from shale. Perhaps this also meant that the US had lower global warming emissions because less CO_2 must have been emitted (though we're not sure about methane emissions as you saw from the last section).

Do we want this switch from coal to gas? Are there other countries similarly addicted to burning coal for electricity? In fact the US is far from being the most coal-addicted nation. There are far worse and these tend to be countries with a lot of coal already and those that have a lot of growing to do. India, for example, will become the world's second largest consumer of coal (after China) by around 2025, with demand almost doubling to 880 million tonnes per year by 2035 (Fig. 7.4). Most of that coal will likely be used to generate electricity to fuel India's economy and thereby pull millions of Indians out of poverty.

The biggest coal user by far is China. It uses about half of the world's coal production and this will increase. According to the IEA, China's coal demand will increase to over 2850 million tonnes per year by 2020 and stabilise above 2800 million tonnes until 2035 (Fig. 7.5) – and this coal will provide half of China's electricity until 2035.

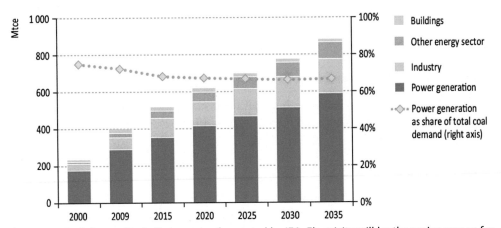

Figure 7.4 Coal demand in India by sector forecasted by IEA. Electricity will be the major user as far ahead as 2035. *World Energy Outlook 2011 © OECD/IEA, 2011, Fig. 10.23, p. 388.*

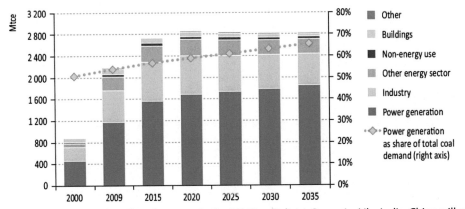

Figure 7.5 Coal demand in China by sector in the IEA New Policies Scenario. Like India, China will use most of its coal to generate electricity. *World Energy Outlook 2011 © OECD/IEA, 2011, Fig. 10.18, p. 382.*

Across the world, global coal use is expected to rise through the early 2020s and then remain broadly flat through to 2035. The scary thing is that we know coal is bad but we're burning more and more of it every year. It will probably remain the backbone of electricity generation unless something replaces it. That something will have to be cheap and plentiful – and for people concerned with climate it will have to have lower emissions than coal. Could the something be shale gas?

The report for the British Government by MacKay and Stone that I mentioned earlier compared flowback in different studies but also went on to compare the overall carbon footprint of different fuels for generating electricity. The authors made assumptions about the size of emissions related to transport of fuel to and from power stations and emissions related to extraction in a life cycle analysis (LCA). For flowback they regarded Howarth's estimates based on the Haynesville shale as 'unrealistically high'.

The result of their comparison is shown in Fig. 7.6.

Calculating lifecycle carbon emissions is not an exact science so results for the carbon footprints of different activities are often presented as a range rather than a single figure. The range of each carbon footprint is shown by the length of the blue column for each fuel when it's burnt for electricity. The position of the blue column above the base of the graph, shows the emissions levels. The units used are rather obscure but can be thought of as grams of CO_2 per kilowatt-hour of electricity generated, in other words the amount of CO_2 per unit of electrical energy.

The whole exercise is rather theoretical in relation to shale gas because, of course, Britain has not produced a single kilowatt-hour of electricity from shale gas yet. Nevertheless what immediately pops out is the carbon footprint of coal to generate electricity.

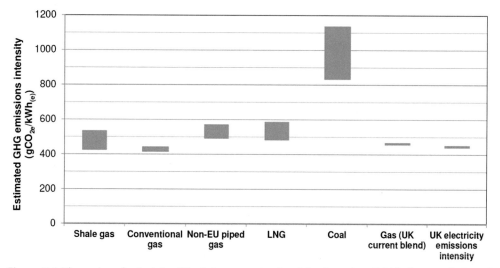

Figure 7.6 The carbon footprints of fuels to generate electricity, based on British figures of LCAs of different fuels. *From MacKay and Stone (2013).*

As the original chemistry suggests it produces about twice as much CO_2 as burning shale gas (or gas generally). But this is partly because MacKay and Stone assumed a low figure for fugitive methane emissions – lower than Howarth would have liked.

Amongst the other things that are interesting is that liquefied natural gas – which in Britain comes from the Middle East – has a slightly higher carbon footprint than would 'home-grown' shale gas. This Middle Eastern gas is conventional (in other words not shale gas) but the fact that it has been pressurised and liquefied and shipped a long way means that the energy put into transporting it makes it more *carbon intensive* than shale gas. That's if the shale gas hasn't been transported very far.

What would it mean if a large country shifted from coal to shale gas? In the first chapter, I mentioned the quandary that Poland finds itself in. It's a big coal user and CO_2 emitter but it doesn't want to be dependent on Russian gas. So using home-grown shale gas might cut its emissions by half *and* increase its energy independence.

But China is a far greater prize. This vast country is already the largest CO_2 emitter on earth. Could it switch to gas? It certainly seems to have a lot of shale gas. China's shale gas reserves could be over 1000 tcf – the largest in the world – in two huge geological basins. China held its first shale gas licensing round in 2011 inviting drilling companies into four areas in the Sichuan basin, and production of gas is expected soon (Fig. 7.7).

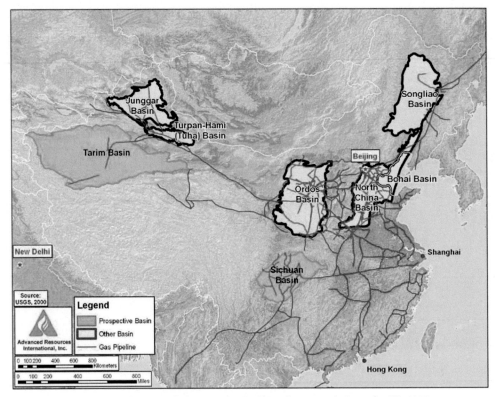

Figure 7.7 The main shale gas areas in China (in orange). *From the EIA, 2013.*

It's not just the industry that promotes shale gas as a bridging fuel. 'Berkeley Earth', a non-profit NGO based in California set up by the renowned climate scientist Richard Muller advocates a switch from coal to shale gas in China through the 'China Shale Fund'. Muller believes that this could be a quick win in the climate battle and perhaps the fastest way to alleviate global warming.

I've not talked about money yet – and I'm no economist – but money obviously has an effect on whether shale gas can substitute for coal, and more importantly can allow a handover to renewables. In fact for it to substitute for coal, shale gas has to be competitive in price in a free market. In simple terms, electricity has to be as cheap generated by shale gas as it is generated by coal. For a big power company to agree to take on gas to generate electricity there must be a general feeling that prices of the raw material will remain stable or at least predictable. The price differences between building different kinds of power stations may also be an issue. For a while now in the US, natural gas prices per unit energy have been lower than for coal. The US has so much shale gas that it could also be self-sufficient in gas for a century and so supply and price stability is good enough to encourage investors in gas-powered electricity.

Gas prices can go too low if there's too much gas produced. As I write the 'Henry Hub price' (a rough proxy for the US natural gas price) is at about $3.50 per MMBTU (million BTU). Some commentators regard $4 as a break-even price for shale gas balancing the high cost of drilling and fracking. At times during the last decade it's therefore been difficult for some companies to make money, if they are producing only methane. Those lucky companies that are able to produce ethane or even oil from shale can increase profits because these commodities are more valuable.

Shale Gas Leading to Renewables?

Wind and solar in the US are all more expensive than natural gas. But this isn't a reasonable comparison anyway because renewables like wind and solar don't provide electricity evenly. The core of supply is the electricity that you can rely upon to always be there when you switch the light or the oven on, or when a factory owner powers up his machine tools or smelting works. This reliable core is known as 'baseload' and it's traditionally provided by fossil fuels or nuclear energy. How to convert intermittent energy sources like wind and solar to baseload is one of the great engineering questions of our age. The problem is how to store energy in huge amounts at times when renewables are producing electricity – so that it can be used when they are not.

For these reasons renewables aren't the choice of most electricity generators. But we need to nurture renewables so that eventually they do become reliable and cheap and can act as baseload. How do we stop power-generating companies and investors – in a free market – from choosing gas every time, so that renewables never get a step up on the ladder?

The answer of course is policy. This is what policy is for – intelligent government stepping in to influence the free market or to address market failures. In the last few decades we've started to value the natural world in purely economic terms because it has a value in its ability to sustain life. By burning fossil fuels we're affecting the natural world's ability to sustain life. When we damage the environment – for example when we put CO_2 in the atmosphere – we make it harder and therefore more expensive to live on the planet. So the use of fossil fuels should reflect that cost, that increased expense.

One of the ways to attach a value to the harm that fossil fuels do is to tax their use. This also has a value in encouraging renewables. Another is to 'cap and trade.' I've already mentioned the European Union Emissions Trading Scheme, which is such a system. The 'cap' is a limit to the amount of CO_2 a factory or a power station can emit. A third method is to use a system whereby electricity suppliers are obliged by government to source an increasing proportion of the electricity they supply from renewables. All of these methods could encourage renewables and ensure that shale gas doesn't outstay its welcome.

DOES SHALE GAS HAVE A PLACE IN MODERN ENERGY?

I suspect that if you started the chapter with a strong view on this question then I won't have changed your mind. In many cases people simply have a gut feeling that shale gas is either wrong or right. There are arguments for and against but it's often an intangible feeling – a gut reaction – that sways your opinion.

For me, the risks from the underground are very low. Earthquakes due to fracking are very rare, and so is water contamination. Surface activities are more risky. With lots of fracking there will inevitably be accidents – spills and surface contamination. The business of fracking at the surface is noisy and will increase traffic, and the landscape will look industrialised. These are manageable activities though – the sorts that you're used to seeing on big building sites – and we're good at controlling them and fixing things when they go wrong. But afterwards the scars of activity will remain for a long time. Much of the world has signs of industrial activity. In Britain it's hard to find a landscape not influenced by modern or historical industry – but for some people it will be one industry too many.

Shale gas is a low carbon fuel, at least in relation to coal, simply because it produces less CO_2 than coal when burnt. Studies of fugitive methane emissions from shale gas drilling show that they are a problem but probably not large enough to offset the advantage that shale gas has over coal. These uncontrolled emissions of course also mean loss of profit for companies and so there is a strong financial incentive to reduce them, quite apart from pressures from regulators and governments. If shale gas is so much better than coal, then quickly cutting China's emissions by half as suggested by Berkeley Earth becomes very attractive, especially if it could be done in other countries too. And surely the right policies applied by committed governments will ensure that shale gas won't stifle nascent renewables, preventing them from achieving baseload status?

Well maybe… Some environmental campaigners are sceptical that shale gas can be adopted and dropped so easily when it's served its function. They've likened the shale gas bridging idea to a fad or crash diet. Perhaps people are deluding themselves thinking they've solved the fossil fuel problem when they're really just hiding it. Isn't shale gas just another dirty hydrocarbon? Do we need to make a clean break from fossil fuels now rather than continuing the addiction?

I don't think there's an addiction. If we could get off fossil fuels easily and quickly we would. My worry is that by taking up shale gas we'll create an industry that will become strong – with jobs attached and lives that are dependent. A shale gas industry will grow and become politically powerful – like the coal industry has. After the business gets going, money, jobs, income, tax revenue can't be ignored by ordinary people and governments. Can the genie of shale gas be put back in the lamp once it's done its job?

But this discussion is rather academic for several countries that either already have a shale gas industry for example the US and Canada – or for countries that are taking the first steps with enthusiastic support from their governments.

Whatever Europeans and North Americans think about it, it's likely that shale gas will be developed in other parts of the world. This week (late 2014) the *Economist* magazine described the potential of the Vaca Muerta shale layer in central western Argentina. The layer underlies 30,000 square kilometres and the US EIA estimates that it could produce 16.2 billion barrels of oil and 308 trillion cubic feet of gas. In 2013, China produced 7 billion (not trillion) cubic feet of natural gas from shale which is a small start, but its government wants a much higher proportion of its gas needs from shale in the future.

Shale gas is likely to be a big part of energy supply in some countries. How can companies be allowed to extract gas, while at the same time protecting the public and the environment? The answer is through regulation – and this is the topic of the next chapter.

BIBLIOGRAPHY

Allen, D.T. and thirteen colleagues 2013. Measurements of methane emissions at natural gas production sites in the United States. Proc. Natl. Acad. Sci. 110, 17768 – 17773. www.pnas.org/cgi/doi/10.1073/pnas.1304880110.

Cathles, L., Brown, L., Taam, M., Hunter, A., 2012. A commentary on the greenhouse-gas footprint of natural gas in shale formations. Clim. Change 113, 525–535.

Energy Information Administration (EIA), 2013. Technically Recoverable Shale Oil and Shale Gas Resources: An Assessment of 137 Shale Formations in 41 Countries outside the United States.

Howarth, R., Santoro, R., Ingraffea, A., 2011. Methane and the greenhouse-gas footprint of natural gas from shale formations. Clim. Change 106, 679–690.

IEA, 2011. World Energy Outlook. (Paris).

McKay, D.J.C., Stone, T.J., September 2013. Potential Greenhouse Gas Emissions Associated with Shale Gas Extraction and Use. Report for the UK Department of Energy and Climate Change.

Miller, S. and fourteen colleagues 2013. Anthropogenic emissions of methane in the United States. Proc. Natl. Acad. Sci. 110, 20018–20022. www.pnas.org/cgi/doi/10.1073/pnas.1314392110.

Pacala, S., Socolow, R., 2004. Stabilization wedges: solving the climate problem for the next 50 years with current technologies. Science 305, 968–972.

CHAPTER 8

Keeping Watch

Contents

Many human activities are controlled through regulation and licences. Regulations allow activities to happen, while at the same time protecting the environment, property and people. Licences are given out to people and organisations with the understanding that they'll abide by the rules. I have a licence to drive because I have to get to work and travel around, but one of the conditions of that licence is that I drive carefully, putting no one in danger. I've passed a driving test to show that I can drive safely. Drilling a well to extract water is also licensed in many countries. Governments know that people need to drill wells for agriculture or domestic use, but they also know that wells can't be allowed to extract more than an aquifer can sustain. Knowing how to regulate activities of all kinds involves understanding the activities and their consequences thoroughly. A lot of science lies behind good regulation. If it's done properly, it not only allows activities, it also keeps them safe while simultaneously reassuring the public. Regulation should be there in the background quietly doing its job.

But maybe shale gas needs more than this quiet regulation. Around the world, people are concerned about shale gas and fracking – the effect it might have on groundwater, the noise, the effect on house prices. Though shale gas companies might have every official licence they need to go ahead, they may not have a 'social licence' – in other words general approval from local residents. People lying down in front of the trucks or picketing the well site might be enough to stop any drilling. Many people simply don't believe engineers and geologists when they say drilling will be safe – and aren't reassured by regulation. Maybe shale gas needs extra work. This and other uses of the underground worry people, because they can't see what goes on deep down. In this chapter, I describe some of the regulation and licences relating to shale gas – and propose a solution to reassuring the public and making people more at home with our use of the subsurface.

Keywords: Baseline; Fracking; Geothermal; Licence; Mineral rights; Regulation; Royalty; Social licence.

This book is about the risk and reward in shale gas. The reward is jobs and growth – and maybe cheap energy. The risk is damage to the environment and human health. In

Shale gas and fracking
http://dx.doi.org/10.1016/B978-0-12-801606-0.00008-X

countries where shale gas is being developed, how is this balance between risk and reward being struck? The answer is mainly through regulation. Regulation can't be too stringent such that it completely stifles the ability for a company or a driller to try different techniques – but at the same time it can't be too lax, such that it doesn't completely protect the public and the environment.

Let's look at the US. In a comprehensive study, a team of lawyers lead by Nathan Richardson of the University of South Carolina looked at regulation across 31 states, considering the types of regulations, how many are used, and how stringent they are. Amongst the 25 main regulations were rules that covered how deep casing should be installed in a well, rules about venting or flaring of excess gas, as well as overall bans and moratoria on shale gas.

The study found that shale gas regulation differs widely across the states. Historically, states have been the main regulators of oil and gas, and since the shale gas boom, states have updated and improved their regulations as and when they needed, taking into account different geology, different water conditions and even different politics. The result is a bewildering patchwork. If we take rules about depth to which casing should be installed in a well for example, most states require casing or cementing at a certain depth below the base of freshwater aquifers, but this varies between 120 and 30 feet (Fig. 8.1).

The University of South Carolina group found predictably that the states with the most gas wells (Texas, Oklahoma, Ohio, Pennsylvania and West Virginia) have a wider range of regulations than do states with less. Interestingly the states with most wells don't always have the most stringent regulations – Ohio and Oklahoma for example seem to be less strictly regulated than many other states with a much less well-developed shale gas industry.

Richardson's group was most interested in what they call heterogeneity – the fact that there is such a variety of regulation across the US. This they attributed to variation in geology, hydrology and population density. In most cases the stringency of regulation matches some need, for example states that use groundwater most tend to have more stringent regulation in that area.

Perhaps the most controversial regulation relates to the disclosure of the contents of frack fluid – in other words whether a company is required by law to reveal the substances it uses in its frack fluid. In a diagram in a previous chapter (Fig. 3.5), I showed the typical mix, but it may be tiny amounts of special chemicals that make a particular frack fluid 'recipe' effective. These additions could uniquely provide a particular viscosity for the fluid at particular pressures in a particular shale layer. It might mean the difference between low and high yields of gas from the shale, so the added ingredient could be a real innovation that could make a lot of money. As in any other area of industry, a special innovation gives one company a competitive advantage over others. So it's not surprising

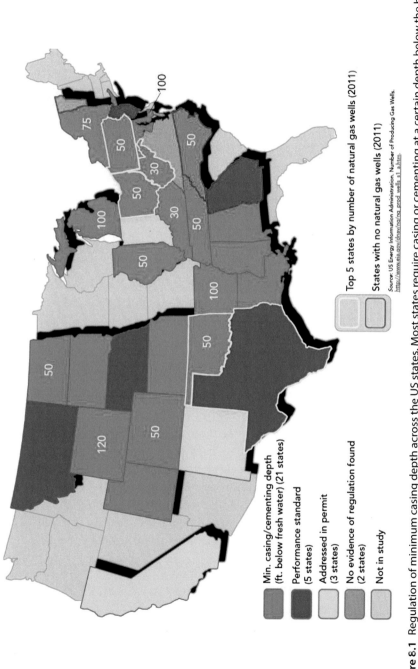

Figure 8.1 Regulation of minimum casing depth across the US states. Most states require casing or cementing at a certain depth below the base of freshwater aquifers. The depth in feet is shown for each state. *From Richardson et al. (2013).*

that companies that innovate in this way aren't keen on telling everyone else what they've discovered.

But it's also not surprising that people who live close to a fracking site would like to know exactly what's being injected into the rocks deep below their farms and houses. It's not very reassuring to be told that any chemicals are harmless, but in the same breath that they aren't allowed to know the names of the chemicals. In the US, the federal Safe Drinking Water Act (SDWA) would have effectively required the disclosure of fracturing fluid until 2005 – about the time that shale gas took off. But in that year the US Congress amended the SDWA to exclude fracturing fluids other than diesel fuel (yes – occasionally diesel has been used as a frack fluid!).

Of the states surveyed by the University of South Carolina group, just under half (14) currently require some form of fracturing fluid disclosure but the detail varies across states (Fig. 8.2). Environmental groups (and some in industry) have suggested that states should insist on disclosure whether the federal government asks for it or not. Many companies voluntarily disclose information on the Web database FracFocus (http://fracfocus.org/).

WHO OWNS THE GAS?

Obviously it's important to know who owns the gas inside shale and who stands to gain by getting the gas out. Legally, shale gas is classified as a mineral, and rights and ownership are covered under 'Mineral Rights' laws.

In most countries the owner of the surface land – be it a house or farmland – has few rights to minerals beneath the surface. In Britain for example, the Crown (the 'State') owns minerals such as oil, gas, coal, gold and silver onshore and offshore. To allow access to minerals like oil and gas, governments like those in Britain and Poland divide up the land into areas known as licence blocks and offer them to companies for exploration and extraction, for a fee. Usually the company also has to commit to drilling a certain number of wells and other scientific analysis. In return the company can explore the licence block typically for five to ten years. If oil and gas is discovered in commercial quantities, then the company has to show to the government how it will develop the field. The plan might include environmental impact assessments and applications for planning permissions for buildings. If the plan is agreed, production can get going and is usually licensed for a set period of time, say 20 years. During this period the company will pay tax to the government.

In the US, however, the owner of the surface land often also has the right to extract minerals from underneath that land. In other words, private individuals own much of the mineral rights across the US. This means that there is a rather different system to commercialise a mineral. In oil and gas for example, the companies contact the owner of the mineral rights directly and negotiate terms. Usually the company

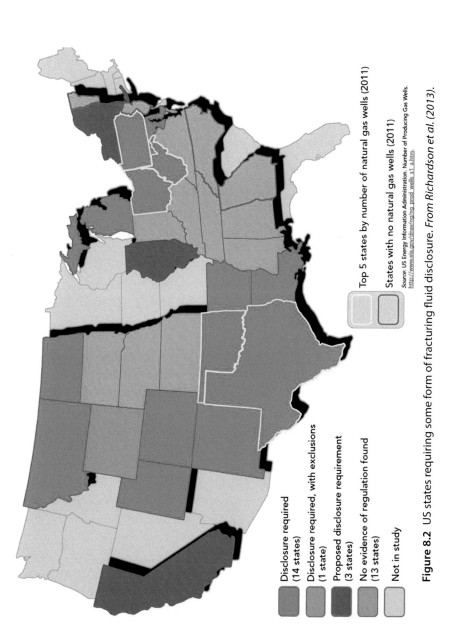

Figure 8.2 US states requiring some form of fracturing fluid disclosure. *From Richardson et al. (2013).*

Disclosure required
(14 states)

Disclosure required, with exclusions
(1 state)

Proposed disclosure requirement
(3 states)

No evidence of regulation found
(13 states)

Not in study

Top 5 states by number of natural gas wells (2011)

States with no natural gas wells (2011)

Source: US Energy Information Administration. Number of Producing Gas Wells.
http://www.eia.gov/dnav/ng/ng_prod_wells_s1_a.htm.

will arrange to lease the mineral rights from the owner and pay a cash sum to the owner. A royalty share of the oil and gas extracted will also be paid to the owner. To ensure that the company does something with the land rather than sit on it, there may also be an agreement whereby the company commits to drill a certain number of wells.

SOCIAL LICENCE

For companies that have government licences to explore and drill for shale gas you might think it's a simple thing to go out and make money. In fact government licences don't mean much if people are lying down in the road in front of frack trucks – or if protestors are blockading a well site.

Despite having a well-developed regulatory and licensing system for oil and gas and several licence blocks that are in good shale areas, Britain has seen very little shale gas exploration and drilling. Even where exploration licences and planning permissions are in place, protests by local people have made operations difficult, or have scared companies off. Sociologists and economists refer to general acceptance on the part of the public that a new technology can and should proceed, as *social licence*. It's quite unofficial and has no status in law. It's often a lot harder to get than official licences, but not having a social licence can stop things happening.

What is it beyond good regulation that reassures people that new technology can be beneficial and safe? I think the key is for the authorities to be transparent in the way that they manage the environment in mining, oil and gas. Let's look at the work of the Environment and Sustainable Resource Development department of the Alberta Government in Canada. Recently it has set up the 'Oil Sands Information Portal'. For those that haven't heard of oil sands mining, this is just about the ugliest way of extracting oil that could be imagined. The thick dark oil with a consistency of treacle is found naturally mixed with soft sand. The sand is quarried at the surface, and when it's deeper underground, it's extracted by injecting hot steam and pumping out the oil. The quarrying and the separation of oil and sand need a lot of power and therefore involve high CO_2 emissions. These processes use a lot of freshwater and produce a lot of contaminated waste and water, which are stored in tailings ponds.

The oil sands are also a huge resource of oil – in fact one of the biggest on Earth – and are obviously very valuable for Alberta. But many Albertans and other Canadians are uneasy about the effects of mining on the environment. The 'Oil Sands Information Portal' provides environmental information on the Web, as far as possible in real time, to show that processes of extraction are not harming the environment. You can log on to a map of the oil sands area, check the locations of all the mines and tailings ponds and

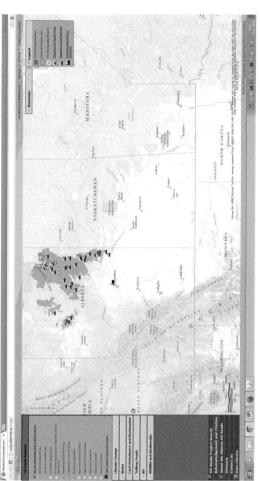

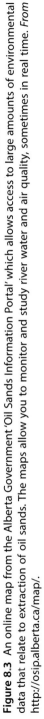

Figure 8.3 An online map from the Alberta Government 'Oil Sands Information Portal' which allows access to large amounts of environmental data that relate to extraction of oil sands. The maps allow you to monitor and study river water and air quality, sometimes in real time. *From* http://osip.alberta.ca/map/.

monitor up-to-the-minute information on air and water quality (Fig. 8.3). The portal allows people to see what's going with the aim of reassuring them that the environment is being protected.

This transparent approach is important beyond the oil sands of Alberta. It may be that for shale gas to gain public acceptance – a 'social licence' – similar environmental data (including from the deep subsurface) will have to be provided freely and openly. But perhaps it needs even wider application, after all public confidence in the way the surface and subsurface is 'managed' is important to several low carbon options for energy.

Let me explain by going back to a paper I mentioned in the last chapter – by Pacala and Socolow in the journal *Science*. In this paper the two Princeton scientists estimated the amount of effort needed to get from the 'business as usual' emissions of greenhouse gases, to a 'flat path'. They focused on a number of 'wedges' that added together would do the job (Fig. 7.3).

We've already discussed a 'fuel switch' wedge where 1400 coal power stations are replaced with gas power stations, but carbon dioxide capture and storage (CCS) also figure in Pacala and Socolow's thinking. This involves capturing carbon dioxide from power stations burning coal or any other fossil fuel, and injecting it deep into a rock layer out of harm's way where it can never reach the atmosphere. From previous chapters in this book you'll probably guess that sandstone is the preferred rock type for this injection – because of its high porosity and permeability. Pacala and Socolow suggested that if CO_2 was captured in 200 large coal power stations worldwide and then stored underground, then this would achieve a wedge. Research has shown that there's enough space in deep sandstone layers to take up lots of CO_2. For example 78 gigatonnes or 78,000 million tonnes of CO_2 could be stored in the rocks in offshore Britain. Considering that a typical big coal power station might produce 10 million tonnes of CO_2 per year – that is a lot of storage space. This kind of burial of CO_2 has been done safely and successfully in the rocks under the North Sea, but when it's suggested on land – in rocks under farms or towns – then it quickly causes public alarm. A recent project to dispose of CO_2 from a refinery, in rocks deep under the town of Barendrecht in the Netherlands, was cancelled because of public opposition.

Pacala and Socolow's nuclear wedge is simple: triple the world's nuclear electricity capacity by 2055 and we'll achieve a wedge. But the simplicity ends there. Apart from the unease that people feel about the presence of nuclear power stations in the landscape, there's also the problem about what to do with the nuclear waste. Britain, which has the oldest civil nuclear energy programme, has 60 years' worth of nuclear waste at the surface in special facilities – and will produce more in the future. Like the British Government, most governments around the world agree that the waste can only be disposed of in a deep geological store or 'repository' in the long term because it will remain harmful for many thousands of years into the future. But again there is a lot of public unease about deep disposal such that development is slow and plans are often abandoned.

Geothermal power is a similar story. Though geothermal power doesn't release any greenhouse gases, its extraction – especially for deep high temperatures needed for electricity generation – uses a process very similar to hydraulic fracturing in shale. This kind of deep geothermal extraction has caused earthquakes – and has an image problem like shale and CCS. Large-scale storage of natural gas in the deep subsurface also has an image problem, even though underground gas storage is much safer than surface storage in tanks.

It's not just about reassurance either. In this book I've tried to highlight aspects of shale gas that scientists don't understand. We saw how shale gas wells appear to be leaking in parts of Pennsylvania. It may be that gas is flowing between the outer edge of the casing and the rock. You can see leaks in the tops of the well casing at the surface, and Robert Jackson's team from Duke University have shown statistically significant evidence that water wells near shale gas wells have more methane. This might be because the cement that's being used to bind the rock to the casing isn't bonding properly. The same problem doesn't seem to exist in other shale gas areas, for example in Arkansas. Why? Could real-time monitoring from specially drilled test wells nearby provide the information we need to solve the problem?

The same is true of fugitive emissions. As you saw in the previous chapter, scientists aren't quite sure how much methane escapes in drilling, fracking and flowback. Rather than measure occasionally, why not measure and monitor all the time and make the results available freely? In this way, measuring and monitoring could help show environmental impacts while also helping scientists make shale gas, geothermal or CCS more efficient.

I could go on but I think the point is clear. Though most of our use of the subsurface – extraction of resources and disposal of waste – is well managed and on the whole very safe, activities underground generally have a bad image and the public is not much reassured by scientists, engineers and regulation. This is the case even in subsurface usage that's very well established and very safe, like gas storage.

Maybe we have to turn to initiatives like the 'Oil Sands Information Portal'. But this mainly covers the surface in the Alberta oil sands area, what would a more comprehensive subsurface monitoring system for shale gas look like?

MONITORING FOR CHANGE

The obvious things to measure are changes in seismic activity (earthquakes), and methane in groundwater. But to detect change we have to know what the normal levels are – before any wells are drilled or fracked. This natural level of earthquakes or methane in groundwater might be very low or even zero but we have to know what it is before-hand. This is called a baseline. Anything above a baseline might be important because of changes going on far below the surface. Many countries already have countrywide seismic monitoring networks and so baseline (natural) earthquake activity is quite easy to work out. There have been very few systematic groundwater methane surveys across the world though. An exception is a baseline survey started in Britain in areas that might in the future see shale gas extraction (Fig. 8.4). This will help to quickly identify if wells are leaking into an aquifer.

Though many countries have a national earthquake monitoring system, these are usually designed to pick up big natural earthquakes not the very small ones associated with fracking or underground mine collapse. In fact the British traffic light system that I mentioned in an earlier chapter – which serves as a warning for escalating earthquake activity relating to fracking – would be hard to administer at the moment. This is because much of the present British seismic network can't detect earthquakes of 0.5 ML at the depth that fracking might occur. The network will have to be denser and more sensitive to detect these earth tremors.

There are monitoring systems for groundwater in Britain but only in a few areas – and these don't measure deep groundwater. The British baseline study I described above mainly sampled rather shallow well water.

So although baselines are ready in Britain and are evolving elsewhere, existing monitoring to watch the day-to-day changes in earthquakes and groundwater above the baseline will have to be stepped up. Many new kinds of sensors could also be used to show how fluids flow in the deep underground. Techniques such as electrical resistivity tomography, which measures the electrical resistivity of the ground, can detect how fluids move, and has been used, for example, to show the interaction of saltwater and freshwater in deep aquifers.

To be able to understand what's really happening deep down at fracking levels, measuring devices will have to be put in specially drilled deep boreholes as well as at the surface (Fig. 8.5). Monitoring boreholes could also be fitted with sensitive instruments that measure movement like tilt or distortion, looking for any deep effects of fracking. Satellites using interferometric synthetic aperture radar can look for tiny amounts (millimetres) of ground deformation or subsidence. Satellites can, in the right circumstances, even look for leaking gases at the surface.

All these data can be collected and presented in close to real time. The 'Oil Sands Information Portal' already does this. The diagram (Fig. 8.6) is a screenshot showing close to real-time data on the water level and water flow in the Peace River above the Smokey River Confluence in Alberta. The same kind of information is available for many locations in Northern Alberta.

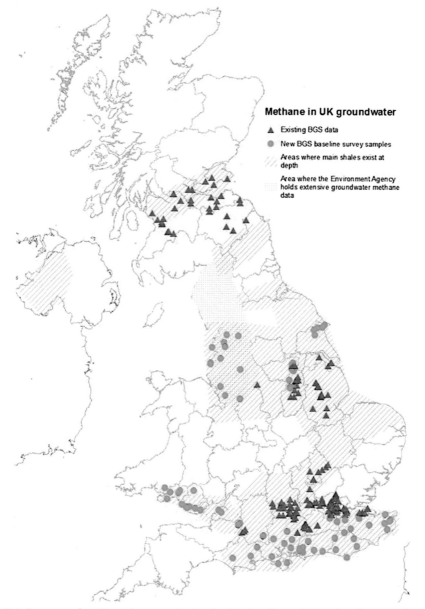

Figure 8.4 A survey of methane in groundwater for Britain will establish what the natural methane levels are before any shale gas extraction. British Geological Survey © NERC 2014. Contains Ordnance Survey data © Crown Copyright and database rights 2014. *From the BGS Website http://www.bgs.ac.uk/ research/groundwater/shaleGas/methaneBaseline/home.html.*

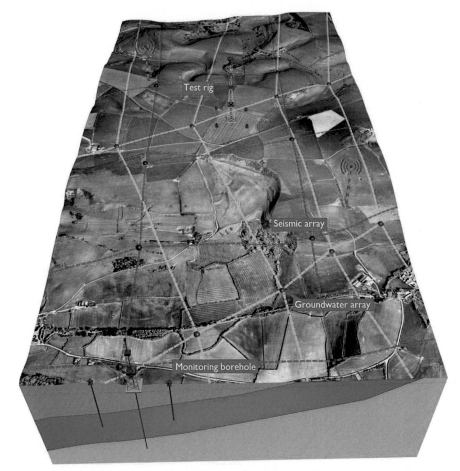

Figure 8.5 To monitor what's happening deep down at fracking levels, measuring devices will have to be put in specially-drilled deep monitoring boreholes as well as at the surface. British Geological Survey © NERC 2014. Contains NEXTMap elevation data from Intermap Technologies. *Diagram courtesy of Chris Wardle.*

IMPROVING THE QUALITY OF THE DEBATE

Even if we do collect data and display them – how do we help people to understand what they're looking at? One of the problems for scientists is not being understood. In the news recently in Britain there have been stories about shale gas and fracking causing volcanic eruptions and enough ground subsidence to allow the sea to flood British coasts. Such things simply won't happen but they have caught the public imagination so that people think they are quite likely. Sadly they have the same 'status' in the media as real concerns such as groundwater contamination. So the quality of the public debate is

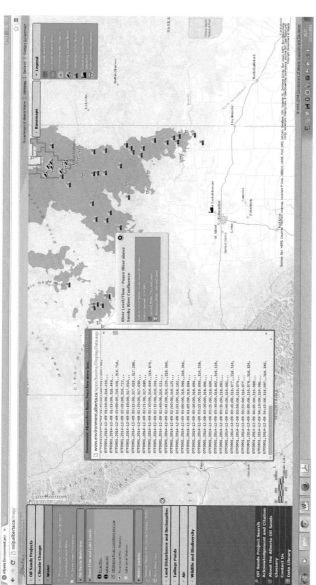

Figure 8.6 Near-to-real-time data on the water level and water flow in the Peace River above the Smokey River Confluence in Alberta. *From the Oil Sands Information Portal http://osip.alberta.ca/map/.*

low. People don't know what really matters and lots of energy is expended worrying about the wrong things. A poor public debate will probably result in poor policy.

Scientists should try harder to explain their work to the people it might affect. But understandably many see their job ending when their peer-reviewed paper is published. Others don't like communicating their work to anyone beyond their specialist field, and in controversial areas like shale gas the temptation to stay below the parapet is particularly strong.

But there are also ways to communicate something of the complexity of science in areas like energy and climate without directly engaging with the public. An excellent example – though not related to geological use of the subsurface – is the British Government 'pathways calculator' which is part of the '2050 Pathways' project (http://2050-calculator-tool.decc.gov.uk/). This project investigates future scenarios to reach an 80% reduction in greenhouse gas emissions by 2050. The 'pathways calculator' is a cut-down version of the large model that's available on the Web. It allows you to build your own scenarios and see the effects. This is important because it shows you the complexity of the British energy system and engages you in developing solutions. It's a kind of 'learning-by-doing'.

Even the wedges of Pacala and Socolow have been turned into a game. The Stabilization Wedge Game was created by the Carbon Mitigation Initiative of Princeton University. It allows you to create your own combination of scenarios to achieve carbon savings, and it's so successful that it's included in several high school curricula. Perhaps similar Web presentations, or even games, could allow the public into the underground world of energy.

However we do it, we have to improve the quality of public and policy debate so that the right decisions about energy can be made. The public should be aware of a range of low-carbon technologies which will help to keep our options open. In shale gas and other geological uses of the underground like geothermal, nuclear waste and carbon dioxide disposal, more transparency and public engagement is needed. Scientists have to talk more about their work. Companies have to be more transparent. Monitoring systems should show how we use the underground. A combination of these measures might reassure the public and improve the debate.

BIBLIOGRAPHY

DECC 'pathways calculator' http://2050-calculator- tool.decc.gov.uk/.
Evans, D., Stephenson, M.H., Shaw, R., 2009. The use of Britain's subsurface. Land Use Policy 134, 34–58.
Oil sands information portal http://osip.alberta.ca/map/.
Richardson, N., Gottlieb, M., Krupnick, A., Wiseman, H., 2013. The State of State Shale Gas Regulation. (Report for Resources for the Future. Washington, DC.)

Stephenson, M.H., 2014. Five unconventional fuels: geology and environment. Unconventional Fossil Fuels: The Next Hydrocarbon Revolution?, 2014, Emirates Centre for Strategic Studies and Research, Abu Dhabi, pp. 13–34.

Stephenson, M.H., 2013. Returning Carbon to Nature: Coal, Carbon Capture, and Storage. Elsevier.

Wiseman, H.J., Gradijan, F., 2011. Regulation of Shale Gas Development, Including Hydraulic Fracturing. University of Tulsa Legal Studies. Research Paper No. 2011-11.

CHAPTER 9

The Science behind the Controversy

Contents

Shale gas is interesting for scientists because it's so full of disputed questions, and because the answers to those questions really matter to ordinary people. There are scientists – astronomers and particle physicists for example – who do very interesting science but have their work cut out justifying its value. In shale gas we're looking at the opposite end of the spectrum – applied science. It's no less difficult or challenging but its findings are crucial for lives and livelihoods because they are concerned with securing enough energy, but doing so without cost to the environment and health. What's great about science is that you can use it to cut up a problem into manageable pieces and then solve each piece. Shale gas is multifaceted but it is amenable to this approach. So in this chapter, I'll weigh up the science evidence and come to some conclusions.

Keywords: Applied science; Coal; Emissions; Fracking; Groundwater contamination; Peer-review; Regulation.

The approach I've taken in this book is to break up shale gas into a set of contested issues – issues that are being argued over. Some of them are shown in Fig. 9.1, but many more could be added.

Just a few minutes of browsing websites, blogs and tweets shows that there's plenty written about these issues. There are also materials in newspapers and magazine articles. It's easy to find websites that say that wells contaminate groundwater – and websites that say the opposite. It's also possible to find complete divergence of opinion on the subject of whether shale gas is lower carbon than coal. But there's no way of knowing whether you can trust the information you read from these sources.

What I've tried to do is concentrate on the best quality information on these disputed issues, which in most cases means peer-reviewed scientific articles. These are articles that have been read and checked by scientists other than the authors and therefore have some claim to be of high quality, and to be independent of influence.

Shale gas and fracking
http://dx.doi.org/10.1016/B978–0–12–801606–0.00009–1

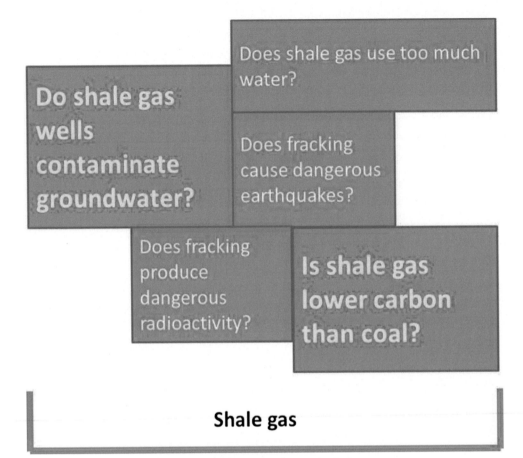

Figure 9.1 Some contested issues in shale gas.

One thing you'll have noticed though is that the peer-review approach, however scientific, doesn't always supply an instant unequivocal answer. In the early days of scientific inquiry in any new subject there will often be several different views, all of which appear to be supported by peer-reviewed papers. This seems a paradox but I think it's a bit like shining a narrow torch beam into a big dark room. You shine a light on aspects of the problem which might even look contradictory. You don't get the big picture immediately.

But that process of shining a light does eventually help you to see what's going on. The business of science – the method used – is instructive too, whether a simple answer is forthcoming or not. The reason is that science journals require that the scientist shows the way he or she tackled the problem. It's a bit like at school doing

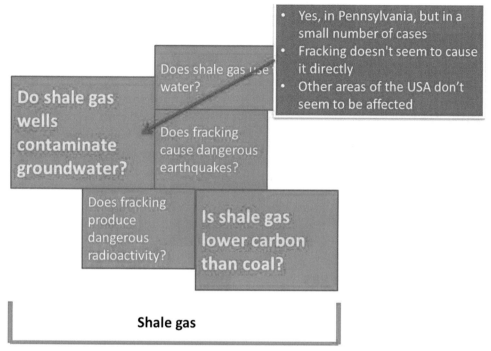

Figure 9.2 Summary of the issue: do shale gas wells contaminate groundwater?

maths problems – you have to show your 'working' and how you got your answer. The 'working' allows anyone with a science training to examine the problem too.

Let's focus on two of the issues, taking contamination of groundwater first. A few years ago it would have been difficult to draw many conclusions at all because so few peer-reviewed papers had been published in the field. But, good opportunistic scientists seek out problems to try out their expertise, and the subject of shale gas and groundwater is no exception. So as of late 2014, there are quite a lot of published papers on groundwater contamination. I've mentioned many of them in this book.

What do these papers say? The balance of evidence seems to suggest three conclusions:

1. There are a small percentage of shale gas wells that leak methane in Pennsylvania;
2. The gas seems to be leaking from the well casing and doesn't seem to be coming directly from fracking deep below the surface; and
3. Other wells in the US that have been extensively fracked for shale gas don't seem to be leaking, for example in Arkansas.

This conclusion is illustrated in Fig. 9.2.

GLOSSARY

Amorphous organic matter Microscopic organic matter in rocks that has no discernible structure.

Anoxia In geology and environmental science, the condition in which environments contain very little or no oxygen.

Aquifer A rock layer containing groundwater.

Baseload electricity The core electricity supply.

Biogenic Formed by biological processes.

British Thermal Unit A unit of energy. See conversion table.

Carbon cycle A cycle that links the carbon in living things, rocks, oceans and atmosphere.

Carbon-intensive Associated with high greenhouse gas emissions.

Carboniferous period A period of time in Earth's history between about 300 and 360 million years ago.

Casing The lining of a well that prevents fluids passing between the well and the rock it passes through, and that prevents the well from caving in.

Conventional hydrocarbons Oil and gas extracted from high-permeability rocks, usually from single discrete geological structures.

Drill cuttings Debris of broken rock from drilling.

Drilling mud A thick fluid used to cool the drill bit and carry drill debris (drill cuttings) out of the hole.

European Union Emissions Trading Scheme A 'cap and trade' trading scheme aimed at reducing greenhouse gas emissions.

Fault A crack or fracture in the Earth along which movement can occur or has occurred.

Flaring Controlled burning of waste gases.

Flowback The waste fluid that returns to the well and the surface after hydraulic fracturing.

Fossil fuel A fuel rich in carbon formed by natural biological processes, followed by geological processes often acting over many millions of years.

Fossil record The evolutionary sequence of fossils through geological time.

Frack fluid The fluid used in hydraulic fracturing.

Fracking Also known as hydraulic fracturing. The fracturing of deep rock using high pressure fluid.

Greenhouse gas footprint The full greenhouse gas emissions associated with an activity.

Greenhouse gases A gas that in the atmosphere allows heat to accumulate, causing the greenhouse effect.

Groundwater Water naturally distributed in rocks underground.

Horizontal well A well drilled vertically (at first) and then turned to horizontal deep down to follow a shale or other rock layer.

Immature In petroleum geology, where a rock containing organic matter has not been heated enough to produce oil and/or gas.

Life cycle assessment A full assessment of the greenhouse gas emissions of an activity.

Liquefied natural gas Natural gas (mainly methane) that has been converted to liquid form for ease of storage or transport.

Mature In petroleum geology, where a rock containing organic matter has been heated enough to produce oil and/or gas.

Methane A fossil fuel with the formula CH_4, which is the most common component of natural gas.

Microseismic monitoring The use of very sensitive seismometers to track the progress of fracking.

Mountain building All the geological processes, chiefly plate tectonic movements, that cause mountains to be formed.

Multistage fracking The practice of separately fracking several parts of the shale layer that is penetrated by a horizontal well.

Natural gas A fossil fuel, usually methane, from rocks.

Peer-review The evaluation of work of one or more scientists by other scientists of similar standing.

Permeability The ability of a rock to allow fluids to flow through it.

Porosity The amount of pore space or void between the constituent particles of a rock.

Produced water Water from the deep subsurface produced as a by-product of oil and gas drilling.

Proppant Small particles, usually of sand, that are injected into new hydraulic fractures to keep them open.

Radon A radioactive gas.

Reserve The proportion of an oil and gas resource that might be possible to extract given economic and environmental limits.

Reservoir rock A rock capable of containing fluids mainly in the pore spaces between its particles.

Resource In petroleum geology this is the amount of oil and/or gas in the rocks. This is distinct from the amount that might be possible to extract given economic and environmental limits – which is the reserve.

Rock cycle A cycle that links the formation of rocks together amongst other earth processes.

Saturated zone The part of the rocks that are saturated with groundwater.

Seismic sections Cross-sections of the subsurface that are constructed using artificial sound waves.

Seismology The scientific study of earthquakes.

Shale oil Oil contained within shale which is geologically mature. This is distinct from 'oil shale' which is shale that contains immature organic matter.

Shale A fine-grained, fissile sedimentary rock.

Source rock A rock that is capable of producing hydrocarbons.

Sweet spots Parts of the shale layer that have high gas content, or character that makes the shale 'frackable'.

Thermogenic Formed by heat.

Unconventional hydrocarbons Oil and gas extracted from low-permeability rocks.

Water table The upper limit of the saturated zone.

Well pad The area around a shale gas well where machinery is positioned.

Conversion Table

Measure	Conversion factor (multiply by)	Measure
Cubic feet of natural gas	0.0001767	Barrels of oil equivalent (boe)
Tonnes of oil equivalent (toe)	7.33	Barrels of oil equivalent (boe)
Barrels of oil (bbl)	0.1589873	Cubic metres
Barrels of oil (bbl)	34.97	UK gallons
Barrels of oil (bbl)	0.136	Tonnes of oil equivalent (toe)
UK gallons	4.545	Litres
Pounds (lb)	0.45359237	Kilograms
Miles	1.609344	Kilometres
Feet (ft)	0.3048	Metres
Square miles	2.589988	Square kilometres
British Thermal Units (BTU)	1055.05585262	Joules (J)
Tonnes of oil equivalent(toe)	41.868	Gigajoules (GJ)
Cubic feet of natural gas	1025	British Thermal Units (BTU)
Kilowatt hours (kWh)	3.6	Megajoules (MJ)

INDEX